PERSPECTIVES IN
Bioethics, Science, and Public Policy

Purdue Studies in Public Policy

PERSPECTIVES IN
Bioethics, Science, and Public Policy

EDITED BY JONATHAN BEEVER AND NICOLAE MORAR

Published in collaboration with the
Global Policy Research Institute
by Purdue University Press
West Lafayette, Indiana

Cataloging-in-Publication data available from the Library of Congress.

ISBN: 978-1-55753-642-6
ISBN (ePDF): 978-1-61249-269-8
ISBN (ePub): 978-1-61249-270-4

Contents

Foreword

This book, *Perspectives in Bioethics, Science, and Public Policy*, is the first in a series of books sponsored by Purdue University's Global Policy Research Institute designed to bring a unique insight into the nexus of public policy and research. The series will appeal to a diverse community of individuals including policy makers, scientists, the general public, and students entering higher education in any field. My observation upon returning to Purdue University in 2010 after serving ten years in the Bush and Obama administrations, most recently as director of the National Science Foundation, is that students being admitted to universities and colleges today have a better developed world view than those admitted a decade ago. I attribute this primarily to access to information on the Internet and greater exploration of contemporary issues in high school classrooms. Many of these students also have had a cultural or learning experience abroad and have a rudimentary grasp of a language other than English. A significant fraction of these students are hungry to learn more about the daunting issues facing the world today, including food security, energy sustainability, availability of potable water, pandemics, invasive species, loss of biodiversity, ecological stresses caused by climate change, loss of biodiversity, and growing shortages of mineral resources. Many of these issues are interconnected, constituting "wicked" problems that challenge even the most powerful supercomputers today.

Some international leaders look to new technologies to mitigate these issues. However, technology alone will be insufficient. These issues also require attention to social (to include moral, ethical, and behavioral factors) and economic factors in developing alternative approaches to solving these problems. These alternatives in turn will need to be subjected to rigorous research and analysis to identify their positive and negative consequences. Policy makers must also consider contextual factors, such as cultural, historical, traditional, environmental, and experiential factors to seek new policies to address these issues that are socially acceptable, economically viable, and ecologically sustainable. Unfortunately, some actions must be taken in the short term and won't wait for painstaking scientific and political consensus building. Calculated risk-taking to identify and test interim approaches and the study of ethical dilemmas associated with these approaches will be needed.

The lecturers in this book understand these matters well and address the moral and ethical issues involved in three categories of contemporary issues, namely, the moral considerations that need to be addressed in using animals for research; the interplay between scientific evidence for climate change and associated policy imperatives; and the challenging ethical issues and policy dilemmas associated with two categories of emerging technologies: nanotechnologies and biotechnologies.

In addition to students, this series of lectures will appeal to scientists, policy makers and the public at large in one very important aspect: they provide excellent examples of outstanding experts in both science and ethics who are able to bridge the communication gap between scientists, the general public, and politicians in discussing challenging ethical issues associated with complex science in a language that is jargon free (or at least jargon defined) and is also compelling, interesting, and laced with humor.

The Global Policy Research Institute is proud to be a sponsor of the Purdue Lectures in Ethics, Policy, and Science series, in partnership with the Office of the Executive Vice President of Academic Affairs and Provost, and academic Deans at Purdue University. This lecture series, which served as the source for this book, was conceived of and led by Jonathan Beever and Nicolae Morar, at that time graduate students. This effort exemplifies a growing interest among students, scientists, and policy makers to apply the Hippocratic Oath "to do no harm" to the conduct of scientific and engineering research.

Beever and Morar have sought to bring to light some of the daunting ethical and moral issues facing scientists and engineers today in identifying research problems to solve and informing the public of the opportunities and risks associated with their research. They are undoubtedly not alone in this endeavor. I expect that students and faculty at other universities are also joining in this movement to help define the high road of scientific research similarly to the establishment by Albert Einstein and Bertram Russell of the Pugwash series of conferences to reset the moral compass of the scientific community in the aftermath of the atomic bombing of Hiroshima and Nagasaki in World War II.

The Global Policy Research Institute is pleased to bring this first in a series of books which will highlight the significant transdisciplinary research that is informing policy makers, thought leaders, students, faculty, and the general public and providing alternatives and consequences to make decisions.

Biographical Sketch

Arden Bement, Jr. retires from his position as the founding Director of the Global Policy Research Institute at Purdue University in 2013. Prior to that position, he was the Director of the National Science Foundation from 2004 to 2010. He served as a member of the U.S. National Commission for UNESCO and as the vice-chair of the Commission's Natural Sciences and Engineering Committee. He is a member of the U.S. National Academy of Engineering, a fellow of the American Academy of Arts and Sciences, and a fellow of the American Association for the Advancement of Science.

Acknowledgments

Sincere thanks to our contributors for sharing their experience and wisdom with us, to the staff at Purdue University Press for helping this idea become a reality, to Ross Blythe for his excellent and professional transcriptions, and to Laurie and Anca for their loving support of us and of our work.

Introduction

This book is grounded in the idea that exploring the intersections of moral beliefs, scientific knowledge, and public policy can enrich our understanding of the value assumptions inherent in bioethical conflicts. Science pushes us to consider the outcomes of knowledge dissemination and the timeliness of our responses to them, ethics guides the normative conclusions we draw from this knowledge, and policy codifies these conclusions into principles for action. This book captures these intersections by bringing together thought leaders from a variety of backgrounds to frame and explore diverse themes in bioethics.

Bioethics, as its prefix suggests, is a domain of philosophical inquiry whose fundamental concern is the value of life. Historically, philosophers have been concerned with analyzing the coherence and consistency of our thoughts about moral value: the right and the wrong, the good and the bad, and the just and the unjust. Similarly, we take bioethics to be the study of what sort of living things we value, and why and to what extent we do so. Unlike traditional philosophical ethics, this work is not done from the armchair. Instead, bioethics relies on a deep partnership with the understanding of the natural world as described by our best scientific knowledge, including not only medicine but also biology, ecology, and the full range of life sciences. Working at the intersections of disciplinary fields and knowledge domains, bioethicists bridge the gap between the sciences and the humanities—two cultures that together can help us apprehend pressing global problems. The conditions of such problems are constantly changing. They cannot be satisfied by clear-cut analytic responses or unique solutions. Hence, they demand constant reevaluation, and for those reasons, they are both the most difficult and most important problems we face.

Bioethics has a fascinating dual origin, beginning in 1970 and developing as a concept and a field in two distinct ways. André Hellegers, Daniel Callahan, and others at Georgetown University helped to structure bioethics around concrete medical dilemmas concerning patients and health professionals (Reich 1995, 20). Two well-respected centers for bioethics resulted from these efforts: the Kennedy Institute of Ethics and the Hastings Center. While this institutional tradition of medical bioethics continues to play a central role in the field of bioethics, it paints only one part of a larger picture. University of Wisconsin cancer researcher Van Rensselaer Potter developed a distinct understanding of bioethics, defending a definition and scope of the concept to account for "a much broader vision" (Potter 1971; 1975, 2300). This vision of bioethics was both evolutionary and ecological, speaking to the moral relationships not only between patients and medical professionals but also between all stakeholders—both human and nonhuman—whose interests may be affected by the outcomes of developing science. This more holistic vision of bioethics, with its scope of long-range environmental and global concerns, now drives a second generation of bioethics.[1] The best of our scientific knowledge, especially from fields like ecology and biology, moves us to reconceive our relations with the world and its inhabitants, while the broader vision of bioethics moves us to reconceive how these relations should matter.

Following in the tradition of public and proactive dialogue in bioethics, the essays here originated as public lectures given at Purdue University between 2007 and 2012. These lectures were designed to bring together scientists, ethicists, and policy makers in conversation. They remain only lightly edited transcriptions of those lectures, rather than fully developed academic papers. They are deliberately presented in a conversational tone to preserve the sense of dialogue sometimes lacking in formal academic discourse. While the following essays aren't typical offerings from the academic world, we consider them a diverse and accessible introduction to some of the central issues in bioethics. The lecture format allows our contributors to tell a story about the difficult moral questions raised by the issues they tackle every day; they have all dedicated their work to the public good, bringing philosophical analysis to bear on real-world issues by advising scientists and by advocating or criticizing policy decisions.

The three sections of this book—focusing on nonhuman animals, the natural environment, and emerging biotechnologies—offer a simple taxonomy of issues that are affected by the complex relationships between moral value, scientific knowledge, and policy decisions. These intersecting relationships have critical importance for every area of bioethics. Within each area, a range of contemporary bioethicists, scientists, and policy makers explore key issues and offer frameworks for thinking about these intersections. They do not, however, offer definitive solutions to the problems discussed, but merely arguments, each open to evaluation, criticism, and revision. These discussions stand as creative building blocks to help us reach the best possible decisions.

In the first section of this book, ethicists come together to consider the bioethical implications of our relationships to nonhuman animals. As our scientific understanding of the nature, function, and ecology of nonhumans continues to develop, the normative consequences of this knowledge are considered and reevaluated. Philosophers Daniel Kelly and Mark Bernstein tell a story about animal minds that reflects this reevaluation. Kelly marks some distinctions between the minds of human and nonhuman animals and suggests the possibility that there might not be empirical evidence that can help in developing policies regarding animal rights. Bernstein argues that the capacity to have a particular kind of consciousness, phenomenal consciousness, is morally relevant—that is, when we deliberate about moral issues, we have to take those individuals with the capacity for phenomenal consciousness into consideration. His arguments have the potential to bring about powerful, practical change, and similar arguments have already led to changes in policy, including a moratorium on research using chimpanzees and the expanding ban on the use of chicken battery cages, and continually increasing scrutiny of the use of animals in research of all kinds.

In his contribution, philosopher and animal scientist Bernard Rollin extends this discussion, claiming that pain is a central and consequential facet of animal experience. The extension of moral value to nonhumans is a fundamental part of this broad vision of bioethics, pushing us to understand and evaluate the moral significance of living things beyond the historical focus on the human animal. This vision of ethics, which takes into account the full range of concerns within and around the life sciences, acknowledges the shifting cultural ethic concerning the impact of nonhuman welfare on human well-being as well as the human impact on nonhuman well-being.

The second section of this book relates the same kind of reasoned moral concern to considerations of the natural environment. With species extinction often cited as the second greatest threat to humanity after thermonuclear war (Takacs 1996, 38) and increases in atmospheric CO_2 levels proceeding at a rate unprecedented in the past 1,300 years ("Climate Change"), the urgency of these issues cannot be overlooked. Environmental philosopher Holmes Rolston III surveys contemporary environmental ethics, arguing that the massive and diverse impact that human animals have on our natural environment creates a need for what he terms an "earth ethic." Likewise concerned with the darker potential of human impact on the environment, philosopher Henry Shue echoes the American poet Emily Dickinson's plaintive worry, "Will there really be a morning?" His lecture underscores the immediacy of global concern over climate change and its implications for the diversity and resilience of all ecosystems. Technological innovation has driven development and economic success worldwide, but technological fixes may not be sufficient to ameliorate the widespread harmful effects of such development on the natural world. Scientist and policy maker Barbara Karn argues for an environmental focus in the development of nanotechnology with an eye on sustainability and limiting environmental impact. As she notes, nanotechnology, which involves manipulation at the atomic level, has incredible potential, but great potential risk.

The third and final section looks toward the future, exploring the role and impact of emerging biotechnologies on our scientific and moral relationships to the natural world. We can see this impact everywhere we turn. The USDA, for example, reports a remarkable 870 percent increase in the use of genetically modified corn, just since 2000 (USDA). Genetically engineered crops, including corn, soy, and cotton, now make up the vast majority of crops produced in the US. From agriculture to medicine to industrial production, what yesterday was merely science fiction is quickly becoming biotechnological reality. Standing at the intersection between the living and the artificial, the authors in this section bring these pressing concerns to our doorstep, from the broadest global impacts of climate change to the unconsidered impacts of the tiniest nanotechnological particles. Nanoscientist James Leary argues that nanotechnology has become more and more relevant to our daily lives. It continues to make inroads, especially in the medical arena. Should we, concerned about ethical implications, constrain nanotechnology's further development to some degree or rather let it progress unfettered?

Biotechnologies have led to controversial collections of information, like those compilations of human genetic data stored in biobanks across the world. Bioethicist Eric M. Meslin responds to this development by examining biobanking's potential risks, for example, concerning the privacy of genetic information or sufficient provisions for informed consent. Such risks in donating genetic information, he argues, are not novel to biobanking and can be minimized by following existing models for donation. Bioethicist Gregory Kaebnick follows with a story of synthetic biology, another controversial and complex biotechnological development. Synthetic biology, which goes beyond genetic engineering and simplifies, modularizes, and standardizes structures to use them in building biological systems, has been called by critics "the final ruination of the world." Despite the call for a moratorium on research in synthetic biology (Pennisi 2012), Kaebnick's essay reminds us that, while developing biotechnologies often pose complex ethical problems, a defensive proactionary stance can minimize risk while allowing science to move forward.

Bioethics, informed as it is by the social contexts of science and public policy, has become a field at the intersections of normative and descriptive inquiry. At these intersections, bioethics brings together philosophers, scientists, and policy makers, and can play a critical role in public dialogue. We hope the essays that follow will help you to critically evaluate your own intuitions about moral value and its relationship to science and policy, to push those intuitions and assumptions in new directions, and to redefine what it means to act morally in your personal and professional lives. Through this book, we hope to enable you to become better leaders, constantly evaluating how best to act for a better tomorrow. Insofar as these essays help you accomplish those goals, they will themselves be valuable as tools in the ongoing processes of bioethics.

Jonathan Beever and Nicolae Morar, February 2013

Note

1. Historian Robert Martensen noted in early 2001 that "approaches more compatible with Potter's expansive definition of bioethics appear increasingly in bioethics journals and forums, if not yet in the leadership of its powerful institutions" (Martensen 2001, 175).

References

Callahan, Daniel. 1973. "Bioethics as a Discipline." *Hasting Center Studies* 1 (1): 66–73. doi:10.2307/3527474.

Churchman, C. West. 1967. "Wicked Problems." Guest Editorial. *Management Science* 14 (4): B141–B142.

"Climate Change: How Do We Know?" Global Climate Change, National Aeronautics and Space Administration. Accessed January 7, 2012. http://climate.nasa.gov/evidence/.

Dickinson, Emily. (1859) 1924. "Will there really be a Morning?" In *The Complete Poems of Emily Dickinson*. Boston: Little, Brown, and Company.

Martensen, Robert. 2001. "The History of Bioethics: An Essay Review." *Journal of the History of Medicine and Allied Sciences* 56 (2): 168–75.

Pennisi, Elizabeth. 2012. "111 Organizations Call for Synthetic Biology Moratorium." *ScienceInsider.* March 13. http://news.sciencemag.org/scienceinsider/2012/03/111-organizations-call-for-synth.html.

Petersen, Thomas, and Jesper Ryberg. 2007. *Normative Ethics: 5 Questions.* London: Automatic Press.

Potter, Van Rensselaer. 1971. *Bioethics: Bridge to the Future.* Englewood Cliffs, NJ: Prentice-Hall.

Potter, Van Rensselaer. 1975. "Humility with Responsibility—A Bioethic for Oncologists: Presidential Address." *Cancer Research* 35 (9): 2297–306.

Reich, Warren Thomas. 1995. "The Word 'Bioethics': The Struggle Over its Earliest Meanings." *Kennedy Institute of Ethics Journal* 5 (1): 19–34. doi:10.1353/ken.0.0143.

Sommers, Tamler. 2009. *A Very Bad Wizard: Morality Behind the Curtain.* London: McSweeney's.

Takacs, David. 1996. *The Idea of Biodiversity: Philosophies of Paradise.* Baltimore, MD: The Johns Hopkins University Press.

USDA Economic Research Service. 2012. "Data Set: Genetically Engineered Varieties of Corn, Upland Cotton, and Soybeans, by State and for the United States, 2000–12." *United States Department of Agriculture.* http://www.ers.usda.gov/data-products/adoption-of-genetically-engineered-crops-in-the-us.aspx.

Biographical Sketches

Jonathan Beever, http://www.jonathan.beever.org, received his PhD from Purdue University's Department of Philosophy in December 2012 and is currently a National Science Foundation postdoctoral researcher in biomedical engineering at Purdue, studying issues of ethics in science. He is the cofounder of the Purdue Lectures in Ethics, Policy, and Science. Beever's work in bioethics as it relates to nonhuman, environmental, and policy concerns has implications for contemporary continental philosophy, political philosophy, and semiotics. He has published and lectured widely on topics including ethics and biotechnologies, ethics pedagogy, biosemiotics, environmental ethics, and postmodern environmental politics.

Nicolae Morar, http://pages.uoregon.edu/nmorar, who received his doctorate from Purdue University in August 2011, is a faculty fellow in the Department of Philosophy and in the Environmental Studies Program at University of Oregon. His dissertation provides an analysis of the ways in which current biotechnologies are altering traditional conceptions of human nature. Morar is a cofounder of the Purdue Lectures in Ethics, Policy, and Science, and has several ongoing projects concerning the role of biology and ecology in applied ethics, the role of emotions in our society, and the role of political power in controlling life.

I. ANIMALS

Our moral world has been almost entirely driven by a human-centered view that has consistently emphasized some set of properties that made the human being unique with respect to the animal world. From Aristotle to Aquinas to Descartes, and from Hobbes up through John Rawls, our moral community was conceived as a function of our humanity, either in reference to our presumed uniquely linguistic character or the complexity of our rational minds. We were not merely different than the rest of the animal kingdom, but our uniqueness was as a source of specialness, of moral worthiness. For this reason, the first generation of thought in animal ethics had to do with whether or not we should accord nonhuman animals moral considerability in the first place. The second generation has to do with the details: by what criteria is this value assessed? How do we rightly choose between two actions, when each has important ethical implications?

The development of answers to such questions, slowly healing the historical rift between the human and the nonhuman animal, has direct policy implications. Clearer scientific understanding helps to support and develop both ethical arguments and policy decisions related to our treatment of nonhumans. Thus, the famous words of British philosopher Jeremy Bentham seem to resonate with us today more than ever: "The question is not, Can they reason? nor, Can they talk? but, Can they suffer?" (Bentham 1907) This statement represents one of the most significant shifts in the history of philosophical ethics concerning how we understand the implications of our relationships to the animal world.

From the use of animal subjects in research to the treatment of animals throughout our food supply chain, legislation and regulation rely on both scientific evidence and ethical framing to guide action and enact change. In this first section, two key animal ethicists and a philosopher bring such evidence and framing to bear on the relation between pain, sentience, phenomenal consciousness, and moral consideration. Their shared working assumption implies that if something is part of our moral community, we have an obligation to consider the ways in which its well-being would be positively or adversely impacted by our behavior. In the case of animals, surely there is some recognition of being better or worse off. The following perspectives help us understand what being better and worse off might be like for nonhuman animals—and by relation for human animals as well. The conclusions of these arguments can leave no future policy maker indifferent.

References

Bentham, Jeremy. 1907. *An Introduction to the Principles of Morals and Legislation.* Oxford: Clarendon Press.

Minding Animals
(2011)

DANIEL KELLY AND MARK BERNSTEIN

Part I: Moral Considerability and Consciousness
Daniel Kelly

Moral Agents, Moral Patients

I want to start off with a view that I think is fairly widely held and fairly intuitive. It's the idea that we don't tend to think of a lot of nonhuman creatures as being moral agents, at least in the full sense that humans are. What moral agency amounts to is still much debated in moral theory and moral philosophy, but we have certain markers that we can point to. We tend not to think of things as moral agents if we don't hold them morally responsible for their behaviors. We certainly don't do so with cows and pigs. We don't think of chickens or crows as being considering or being bound by moral duties or obligations either, and I certainly don't think of cats or dogs as doing anything like deliberating—let alone deliberating about whether some behavior is virtuous or vicious—before they engage in it. I will just grant this as an assumed premise for our purposes here. It sounds plausible enough, but granting it doesn't exhaust the kinds of questions you can ask about the relationship between animals and morality. You can also separate out this question of moral agency from the question of what it takes for a creature to be a moral patient.

What do we have in mind when we talk about moral patienthood? It's analogous to the question, "Are there nonhuman creatures (and I want to use creatures in an open ended sense) who deserve our moral consideration? Should they be subjects of our moral concern?" We can break this down a little bit further, asking, "Are there living entities that should be treated ethically, entities whom we should consult our moral theories about when we engage in a behavior that directly impacts them?" Another way to say this—Peter Singer is famous for this piece of terminology—is, "Should other nonhuman creatures be considered part of our moral circle?" If we grant that some other nonhuman creatures should be morally considerable, another questions arises as to which ones. Is it all living creatures? Should not only the dolphins and maybe the chimpanzees, but also the evergreen trees and the redwood forests, and paramecia all be part of the moral circle? Or, maybe it's just higher order animals? For instance, maybe deer get included there and maybe tarantulas do, but the viruses don't. Do the bacteria? Dolphins, maybe bats, but maybe not shrimp? These sorts of questions demand answers.

If we buy into the idea that there is some dividing line between the morally considerable and everything else, we can ask, "Well, what is it that some creatures have that other creatures lack that qualifies them to be in the moral circle? Why is it that they deserve our moral concern, and that other creatures don't?" One philosophically precise way to ask these questions is, "what are the necessary and sufficient conditions that need to be satisfied for some creature to be a moral patient, to be part of our human moral circle?" Then, setting that question aside, we can also separate out another question: "if some nonhuman creature gets in the moral circle, how much moral consideration does it deserve? Should it be treated as equal to humans, or should it get just a percentage of the consideration humans get?"

Mark Bernstein will give arguments in favor of broad moral consideration for nonhuman animals in the second part of this chapter. My goal in this first part is to get us to a place to talk about some issues concerning mind and consciousness where we can say with a little bit more precision what the necessary and sufficient conditions might be for moral patienthood.

Continuity Between Human and Nonhuman Minds

Let's start with similarities and differences between human minds on the one hand and nonhuman minds on the other hand. A fair question you might ask is: "Why is this relevant? If the question is one of moral concern and moral considerability, why do these differences and similarities matter?" The reason I'm going to discuss these issues a little bit is that they are also relevant to this question of moral consideration. We can factor out the different features and capacities that different minds might have which might be directly relevant to this question. The mind is a complicated thing. We can be careful about its different features and capacities, and think about this as separating out the morally relevant wheat from maybe the merely psychological or physiological chaff. What I'm interested in has to do with the evolution of human cognition and the features of human cognition which make human minds distinct in the animal kingdom and which make our minds unique.

Nothing I'm going to say here is uncontroversial; a lot of work is still being done on the boundaries of what we think is distinctly human about our minds and what is shared with other minds. One of the first places to go in terms of a straightforward cognitive capacity or feature of human minds that might be unique is just our ability to easily acquire and then use complicated language: not just rudimentary language, but, roughly something that has a fairly complicated syntax and involves recursion (the ability to take one phrase and embed it in another phrase, and embed it again in another phrase). The human capacity for written language appears to extend and elaborate on this capacity for spoken language. So unlike other creatures, we have artifacts like the Code of Hammurabi, and the complete works of William Shakespeare.

A lot of interesting work is being done right now on what you can think of as perhaps a second unique ability of human beings: the human capacity for social learning and how our sophisticated capacities for sociality plug into the fact that we are extremely cultural creatures—almost uniquely cultural creatures in the animal kingdom—whose cultural knowledge can accumulate from one generation to the next. An important element of this is called *observational learning*, which allows humans to watch one human perform some behavior and then learn how to do it through such observation and imitation. And although this, too, isn't totally uncontroversial, it looks like a key component of our ability to accumulate cultural information over generations. Related to this is the observation that human minds, brains, and psychological capacities are extremely flexible. Sometimes this is talked about in terms of *behavioral plasticity*. This feature is part of what has allowed humans to survive, in fact, to thrive in a number of different habitats, to an extent that is quite uncommon in the biological world. Humans can live in the desert, live in the tundra, live in rainforests, and thrive in the mountains. We have *behavioral flexibility*,

which is connected to this ability to learn socially and to accumulate knowledge about the environment from one generation to the next. Many other human capacities might be shaped by these sorts of considerations (see Richerson
and Boyd 2005; Sperber 1996).

A third and perhaps most important and interesting attribute is the capacity that humans have to live in large cooperative groups. Psychologically
speaking, it looks like we have a capacity to acquire and then, once we've in-
ternalized them, to comply with behavior-guiding rules. Think of these rules
as *social norms*. It also might be the case that once we've acquired a norm or
rule, we are motivated to punish those people we see violate that particular
norm. Another component of this package is of capacities are 'tribal instincts',
as some of theorists call them. These have to do with our sensitivity to group
memberships or to what are sometimes called *tribal markers* or *tribal bound-
ary markers* (see also Henrich & Henrich 2007).

Some other features that theorists have thought were distinctive to human minds have a more qualitative or affective component, such as certain emotions that might be uniquely human. Some of these theorists have
thought there may be not just uniquely human, but even some culturally
specific emotions. Sometimes these are called culturally bound syndromes
because they are partially socially constructed, and different societies can
construct different emotions. One that people have heard of is *schadenfreude*,
which started in Germany. Schadenfreude is a feeling of pleasure, but it's
a very particular kind of pleasure. It has to be triggered by someone else's
suffering. But there are other ones you can read about in books and anthro-
pological journals, including one called *latah*, which is explored in a nice
book called *Boo* (Simons 1996). Latah is a cultural syndrome that you find
in Indonesia, typically experienced by women. It's triggered by a startle. If
you startle a woman who is subject to latah she goes into sort of a trance
state where she is also extremely suggestive. Sometimes she will engage in
automatic-seeming behaviors and repetitious speech. One interesting thing
is the way this particular emotion is also woven into the norms of this society—what a woman is who is experiencing latah does, she is not held morally
responsible for. She is off the hook, so to speak, for whatever happens to her
after the fact.

Then, there are some other emotions that might not be culturally specific
or socially constructed, but there is a good case to be made that they are in
fact uniquely human. Emotions like guilt, shame, and pride are getting a
little bit closer to emotions that are important for morality—likewise for the

emotion of elevation. I tend to think of elevation as the feel good emotion. This is an emotion that has been investigated by positive psychologists recently. It's triggered by witnessing some act of moral virtue or some display of moral beauty. It gives you that feel good feeling, a sort of emotional dilation. It gives you motivation to go out and do something good yourself. John Haidt talks about it as the emotion you most associate with a really good episode of Oprah (2006).

Again, none of this brief summary is uncontroversial. It was merely a quick run through of the sorts of things that cognitive scientists have thought might be distinctive features of human minds. Now we can ask the question, a bioethics question: "So what? What does any of that have to do with morality?" As you saw in the beginning of this chapter, this question actually fragments into two questions. First, are any of these capacities necessary and sufficient or essential to moral agency? And second, are any of these capacities necessary or sufficient to moral patienthood? Are they required to be the proper or deserving creatures of our moral consideration?

My suspicion is that this area of empirical moral psychology, which is on the rise right now, will provide us down the line with a better understanding of many morally relevant psychological capacities, and maybe some of those psychological capacities that I just mentioned will turn out to be agential capacities. Investigating them can eventually help us to better understand how the image of humans delivered by science might connect up with our lived experience of ourselves as persons and moral agents.

But our question for today isn't about agency; it's about patienthood and about moral considerability. So we might also approach this question by asking about the features of human minds that are shared with other animals. How is the human mind continuous with other cognitive systems, other brains in nature? There is a long tradition of asking similar questions that goes back at least to Darwin. Darwin thought there were important breaks between humans and other creatures in terms of their psychological capacities, but there wasn't a fine cut or an absolute divide. Instead, he thought that you can find a lot of nascent capacities in animals that we find in more well-developed or conditioned humans.

The sorts of cognitive capacities that we can find in some other creatures, like some of the building blocks of language, seem not to be found in all creatures, so maybe not all creatures have the sort of ease of acquisition of a language that humans have. You can teach Nim Chimsky how to sign and you can teach other kinds of apes how to sign. It requires effort to teach them—it doesn't happen as easily as it does for little baby humans—but you can teach them. Similarly, there has been interesting work done on songbirds (Rothenberg 2005). A case can be made that some songbirds actually use recursion, giving their songs a complex syntactic structure. The rudiments of language, then, can be found in other animals, like birds—suggesting language is not wholly unique to human beings.

For a long time what was thought distinctive about humans was the use of tools. That theory is pretty much shot at this point. Crows use pieces of leather and pieces of leaves to help gather insects and there are now famous examples of chimpanzees of using sticks to "fish" for ants in anthills.

Perhaps the capacity for having a *theory of mind* is crucial for moral considerability? Actually, calling it a "theory of mind" is a bit of a misleading name for such a capacity. But the basic idea is that we humans ascribe beliefs and desires to one another. That's how we understand one another, and we make predictions about what other people are going to do and how we make explanations about why someone did what he or she did. There's a huge body of research about this capacity in the cognitive sciences, and it is in part a comparative project. Do other apes and gorillas have comparable theory of mind capacities? It looks like the capacity is not as sophisticated or full blown as in humans, but maybe some of the rudimentary components can be found in apes and gorillas. Interesting work has shown that, in some respects, dogs are even a little bit closer to humans than apes and gorillas. The driving idea here is that they've been enculturated with humans for so long that dogs now gaze monitor humans—they pay a bit more attention to where humans are looking. This feature of our minds too seems shared with other animals.

These are straightforward cognitive aspects in which human and animal minds might be similar. Other physiological aspects of minds, like complex perceptual systems, can be found all over the animal world. Things like eyes appear to be evolutionary good ideas. What's an eye? It's a way to gain information about the immediate environment via light waves. We find those not only in other mammals but also we find them way far away from us on the tree of life or on the phylogenetic spectrum. Different kinds of eyes, same basic idea. So sophisticated perceptual systems, like ours, can be found fairly widespread in nature.

Some kinds of affective capacities look like they might be found in nonhuman creatures as well. These might appear very much like fear, like the emotion of anger, or maybe very much like the emotion of surprise. You do have to be careful not to anthropomorphize when doing a lot of this work. My suspicion is that with many emotions it's very easy to anthropomorphize. But there is certainly something like fear out there in many other kinds of mammals. And it's plausible that mammals are not the only animals to experience emotions like fear and surprise.

So the upshot is that there is a range of interesting work being done on how human minds are continuous with nonhuman minds. But you can ask the same question, "So what? What does this have to do with the main focal point of our discussion?" Well, we think with this last comparison we are a little bit closer to what is important. I want to narrow down the focus to one key feature. This feature, we think, is important in virtue of its connection to just having a point of view, having a capacity to feel pain or pleasure, and therefore having a capacity to have interests. In a word, what this feature is, of our minds, and even some animal's minds, is consciousness.

Phenomenal Consciousness: Modifiable Hedonic Experience

Discussions of consciousness often remind me of the Tower of Babel. Analytic philosophers of mind tend to have something very specific in mind when they use the word, but consciousness gets used in a variety of different senses; particularly flagrant, differing uses when you go from one discipline to the next to the next. Because of this, conversations about consciousness can be confusing and a bit frustrating. Again, part of the ambiguity of the term comes from a slightly different use in every subdiscipline. There may be a different use in every theorist's mouth. But the one sense of consciousness I want to focus in on is what analytic philosophers of mind call *phenomenal consciousness*.

Phenomenal consciousness is, in a word, experience. It's the raw feeling or qualitative component of any mental state that is phenomenally conscious. Consider this turn of phrase that gets a lot of mileage: a creature is phenomenally conscious if there is something that it's like to be that creature, that is, if that creature has *qualia*. Qualia are the properties of, or the aspects of, those mental states that are phenomenally conscious. They are the parts of the mental state that are accessible or knowable from the inside: the light is on, you can know that from the inside, from the first-person point of view. In other words, qualia describe the sense of consciousness directly linked to subjectivity or a subjective point of view.

When they get pushed, philosophers of mind trying to pin down what they mean in talking about phenomenal consciousness tend to use ostensive definitions. They refer to the qualitative component of a variety of mental states: you might whack your thumb with a hammer and experience pain. But the mental state of pain has a lot more to it than just the sensation. It has typical upstream effects like sensation, surely. But things that cause pain like damage to the body typically have downstream effects, too, like behaviors you might engage in when you're in pain. There are typical downstream cognitive effects as well. If you're in pain, part of what that does is take your attention and draw it to the source of the pain, so you can make it stop. On top of all that there's also just the raw feeling of it—hurting. That's the qualitative aspect, the sort of experiential component of that particular mental state. Philosophers tend to go to pain and talk about that one a lot. But you don't have to just talk about this negative state. You can talk about the nice "rubby" feeling of a foot massage. Or, you can talk about disgust— I do because I know a lot about disgust (Kelly 2011). There is a lot more to disgust than just a flash of nausea or that aversive feeling that you get when you're disgusted, like when you see a nasty diaper. But that's the qualitative, experiential part of disgust. And then you can talk about the qualitative part of perceptions as well. So the redness of a rose, the intrinsic redness of it, is the qualitative aspect of your experience of looking at a rose. Or there's the intrinsic greenness of that light that Gatsby believed in and chased after.

Another way that philosophers of mind tend to creep up on and characterize this is talking about what this kind of consciousness, phenomenal consciousness, isn't. Or what phenomenal consciousness and qualia can't be. There are raging debates about how to explain what qualia are, and involved in that debate are some philosophers who are skeptical about the entire project. They say something like, "Look, I don't understand what you're talking about when you talk about qualia. Set aside the explanations of them, and give me a good characterization of the phenomena themselves." One of the responses to this request is to reply, "Look, qualia is like jazz—if you got to ask, you ain't never gonna get to know" (Block 1978). There is something ineffable about phenomenal consciousness, on this view.

Another way to try to define qualitative consciousness or phenomenal consciousness by negative definition can be drawn out of a very famous argument made by Thomas Nagel. Roughly, the character of their phenomenal consciousness is one thing we can never know about bats. Now, we can know a lot of stuff about bats and I submit that, at this point, we do know a lot of stuff about bats. But imagine yourself ahead twenty or even forty years, when we have complete and exhaustive information about bat physiology, bat anatomy, and bat neurochemistry, everything you want to know. We know how their perceptual systems work. We know how information is gathered, integrated, and then fed back out to drive bat behavior. What Nagel says is that you can know all that, but you still would not have touched the question of what it's like to be the bat (Nagel 1974). Nagel thinks that's still a wide open question; none of the science speaks to it.

Nagel makes this point in a famous paper where he uses the example of a bat to illustrate his argument, a brilliant rhetorical choice. It's one example, but he picks the bat because bats are close enough to humans: they're mammals and they have sophisticated nervous systems. They're close enough to humans that most people think there is, indeed, something that it's like to be a bat, that bats are probably phenomenally conscious, that they have experience. So that's good. On the other hand, bats are pretty alien too. They're mammals, but they get their information about the world through a sensory modality that's nothing like any of other ones we have. Bats navigate by echolocation. They use sonar, basically. Nagel's point is that you might know how echolocation works, and indeed we do know how echolocation works. What Nagel says is you know how echolocation works but you don't know what it's like to experience the world through the sensory modality of echolocation. You still don't know what it's like to be a bat.

The choice of the bat, while brilliant, is also merely rhetorical. The philosophic difficulties that he's pointing to have nothing to do specifically with bat neurology. It doesn't have anything to do specifically with nonhuman or animal minds. What it has to do with is phenomenal consciousness. How can we know what bat qualia are like? How can we be sure that bats are conscious at all? Most people have the intuition that bats are phenomenally conscious. How can we know? There is nothing harder or easier about that question than another question I can ask, namely, the same question about anyone out there. How can I know any of you are phenomenally conscious? How can I know that you have qualitative experience in the same way I do? Maybe the spectrum as you experience it is completely inverted compared to the way I experience it. That's a tough philosophic problem, and it's exactly the same sort as the one raised by Nagel's question about bat consciousness.

We can open up the scope of this question a bit, too. Trying to figure out where phenomenal conscious in any of its instances—bat, human, or otherwise—fits in the natural world is a tough question. We all have an organ encased in our skulls doing lots of complicated stuff. There are electrical signals being sent back and forth via neurons. There are neurochemical things happening that I don't begin to understand. In addition to the chemistry and biology, the brain all gives rise to vivid technicolor experience (see McGinn 1989). But I have a lot of complicated organs in my body doing lots of complicated chemical and biological things, and most of those do their work "in the dark." They don't give rise to anything like experience. What's special about the brain? Where do the qualia that it gives rise to fit in the natural order?

So here is where I hand off to Mark Bernstein, and I end on this point for a couple reasons. The first is that this particular feature of minds, qualia or phenomenal consciousness, is where Mark wants to start the discussion of moral patients, moral concern, and the moral circle. But I also want to acknowledge that these questions about phenomenal consciousness are very tough. Entire literatures, entire subfields of philosophy of mind are devoted to these sorts of questions, and most of the special difficulties with phenomenal consciousness or explaining qualia have nothing to do with the unintelligibility of other animals or the biological boundaries between species. These problems having to do with consciousness and subjectivity are tough problems in metaphysics, but they are just as tough and real for humans as they are for other animals.

Part II: From Minds to Minding
Mark Bernstein

Having given you some idea of what phenomenal consciousness is or what qualia are, I want to now argue that the capacity to have phenomenal consciousness, at least some kinds of phenomenal consciousness, is both necessary and sufficient for being part of our moral domain. That is, the ability to have certain kinds of experience—and I'll tell you what kinds in a moment—is going to be both necessary and sufficient for moral considerability. When we morally deliberate, we're going to have to worry about those individuals, and only those individuals, that have the capacity for this type of phenomenal consciousness. This move is not going to be counterintuitive. The really strange stuff will come at the end.

Let's say the only experience a creature could have is the experience of red. So introspectively it had the experience of red, whether it be a red square or red circle—it doesn't matter. Would that be enough to put it as part of our moral domain, our moral scope? The answer to me seems to be no. Why not? Because if that's the only introspective phenomenal experience it could have then you couldn't make the creature better or worse off. You couldn't affect from the inside, since by definition, by stipulation, the only thing it could have from the inside is this qualia, this experience of red. There would be no way to improve its well-being; it would have no well-being to improve or diminish.

Interest in Survival Matters—Dogs in, Rocks out

Well, then, what is the next step? For an individual to be morally considerable the creature needs to have the capacity for aversive and enjoyable experiences. It has to have the capacity to be made better or worse off from the inside. The typical example is feeling pain. If I hit you over the head with a hammer you're going to feel bad, you're going to hurt from the inside, and in experiencing that pain you're being made worse off. So now, you might think (and almost every philosopher thinks this, although they're wrong as I'll show you in a second), that the capacity for what you might call hedonic experience—the experience of being made better or worse off or, roughly, the experience to feel pleasure and pain—is enough to put you in our moral community.

That's not quite right. Here's why. We can imagine a creature that has the capacity to be made better or worse off from the inside, and yet the experiences that constitute it being made better and worse off cannot be modified by us. Take an example of somebody who experiences interminable pain and there's nothing we can do about it. If that were the case, that individual could not be part of our moral community because we could not make that individual better or worse off from the inside. Now that might seem very theoretical and have no relevance at all. But at least on some conceptions of God, God is immutable. He can't be changed by anything. So if God is able to experience, say, joy and it's true that that experience cannot be changed by anything, then God is not part of our moral community. The conception of unchangeability makes God a person outside of our moral domain.

Now I think we have reached the crux of what moral considerability is. Roughly, it's the capacity to have hedonic modifiable experiences. If an individual has phenomenal consciousness, or qualia if you will, as long as its experience has hedonic quality, so long as that individual can be made better or worse off and that individual can feel pain and pleasure, and that experience is modifiable (that is, we can do something to change that state of mind), then and only then does that individual deserve our moral concern.

Well, if that's right, what does that leave out and what does that put into our moral community? Let me just give you a few examples. I don't think these will be strikingly bizarre to you. Rocks would not be part of our moral community, because I think we all assume—and I hope I don't have to prove this (I don't know if I could prove it)—that rocks have no phenomenal consciousness. They have no inside, they have no interior experiences. You can certainly do things to rocks. You can throw them, you can shatter them, you can throw them at glass windows, but you can't make them better or worse off. They're not phenomenally conscious; nothing matters to them from the inside and so there is nothing you can do to a rock to make it better or worse off. Therefore it can't be part of our moral community. Pieces of paper would correspond to the same idea.

How about computers? Well, I think they're not part of our moral community now, but I think they might be. That is, I don't see any reason to think that in the future—God knows when, twenty years, fifty years, or four hundred years from now—computers that have phenomenal consciousness will be built. They actually will be able to suffer, experientially suffer, and experientially enjoy. And if they do, and if we can do something about those states, then according to my view they deserve to be part of our moral community. We have to be concerned about their welfare or well-being when we deliberate. Again that's not the state of things now. I'm not claiming that computers are feeling things right now, but I don't see why in principle they couldn't in the future.

OK, well, so what's in now? What currently constitutes our moral community? Well, obviously we do; that is, normal, adult human beings. And now the question is, of course, "What about nonhuman animals? Do nonhuman animals have hedonic, modifiable, inner experiences?" It seems to me that the answer is (and I'll give you the reasons in a second) obviously, yes! I'll put this somewhat contentiously: what person in their right mind doesn't think their dog or cat has the capacity to suffer or enjoy things? None of you! No one does except for a few philosophers. When you go to a veterinarian—how many of them do you think believe that animals do not have the capacity to feel pain or pleasure? Zero. Vets give dogs and cats pain medication. What would be the point of giving them pain medication if they didn't feel pain? Now you might say it's to make their owners happy: even though their animals really don't feel pain, the owners think they feel pain, so if we give them the pain medication, then the owners will feel happy but it's really not doing anything for the animals. But if they really did believe that—and none of them do—why bother? Give dogs and cats a placebo, and explain to the owners that they're under a common misimpression. It would be a lot cheaper.

So everyone, virtually everyone, believes that at least some animals have the capacity for these phenomenally conscious experiences. I did not say *all* animals. There are clearly some animals that don't: bacteria, amoeba, and paramecia surely don't and maybe if you go somewhat down in the phylogenetic scale others don't either. I'm not an expert on the science of conscious experience. For example, do oysters? I don't know. Do cockroaches? I don't know. But certainly there is no doubt that when we get to a certain level—certainly mammals and primates, and I would argue strongly for many fish if not all—there is such experience. There has been research on this by Joseph Garner and others that fish have the capacity for these kinds of experiences (Nordgreen et al. 2009). But let's not cavil. Surely dogs, cats, monkeys, cows, chickens and so on all have the capacity for these phenomenal experiences.

Why is this important? Well, for one thing, in the United States alone approximately ten to eleven billion nonhuman animals, primarily pigs, chickens, and cows, are killed for food annually in factory farms. There isn't factory farming of cockroaches, to the best of my knowledge. That's kind of a side issue. Do cockroaches have the capacity for these kinds of experiences? Like I said, I'm not sure. If I have a cockroach in my house, I will shoo the cockroach out. I won't kill it. I err on the conservative side. It doesn't take me that much effort to shoo them out of the house and I do so. But again, I'm certain that dogs, cats, pigs, chickens, and cows do have such a capacity. So if such capacities warrant moral consideration, it obligates great change in things like the practice of factory farming.

Well you might say, "What evidence do you have?" You might think my claim about animal consciousness is purely question begging, I'm just saying they do. What evidence do I have? Well, I think there are three kinds of evidence. There is behavioral evidence; that is, some animals act exactly as if they had phenomenal experience—at least very much like they would act if they were in pain. I mean if you thought a dog felt pain in its foot, how would you expect him to behave? Presumably, he'd lick his foot, withdraw from the cause of the pain and discomfort, and so forth. Actions like that offer behavioral evidence that dogs, cats, et cetera, have these inner experiences. There is also physiological evidence. There are endorphins that get released in human beings as a way of relieving pain. These same kinds of endorphins get released in animals while they're in pain. Coincidence? I don't think so. And there is plenty of other physiological evidence as well. I won't bore you with the details. The third type of evidence, and it goes along with some of the physical evidence, is evolutionary evidence. I guess if you don't believe in evolution then evolutionary evidence won't mean much, but to the huge majority of us, it is important to note that evolution is continuous. It would make sense that animals at least very close to us would have these same kinds of inner sensations, and certainly Darwin thought so.

Now you might say that none of that is conclusive. Even if you say "Yes, there is some behavioral evidence, there is some physiological evidence, there is some evolutionary evidence, but none of that proves beyond a shadow of a doubt that the dog feels pain." I agree! But, as Dan alluded to at the end of his comments, how do I know you're feeling pain? What is my evidence? Your behavior. What other evidence do I have? The only behavioral difference between you and the dog is that you're telling me. You're saying, "Bernstein, I'm in pain" whereas my dog goes "woof," or "bark," or whatever he does. But, on the other hand, how do I know that you're not lying to me? How do I know that those sounds, "I am in pain" reflect really what's going on inside of you? If you really want to play this skeptical game I can play it with you as well. Is it proof, absolute proof, from the fact that you say "I am in pain" that you are in pain? Of course not. Again, you could be lying or you could even be mistaken even about your own pain.

So if there is no proof in the case of the dog or the cow, and that it's not a conclusive argument from behavior and evolution and physiology to animals really feeling, then ok, there is likewise no conclusive evidence for the similar human case. So I would say we have just as good reason in general under ordinary circumstances to believe that a dog is in pain as we do to believe that a human is in pain.

If all of this is right, what conclusion do we draw? Well, first, many, though not all, nonhuman animals deserve concern. Again, for most of you this is common sense. People who have dogs and cats, as I'm sure many of you do, think they suffer sometimes and you do things about it: that's why we have veterinarians. You bring them to the vet and say something to the effect, "My dog's leg is hurting, he's limping." Virtually everyone thinks that.

A more interesting question, I think, is, "How much concern do animals deserve?" OK, you could say the dog deserves some consideration, some concern, but how much? My view is that animals and humans deserve the same consideration, and that, if one disagrees, the burden is to show why humans (presumably) are entitled to greater concern. I don't think this burden can be discharged. But after explaining a bit on what ascribing equal consideration amounts to, I want to disentangle this question of comparative concern from another with which it is often confused, namely, the question of the comparative value of the lives of animals and humans. And now here's my somewhat more startling thesis, if you will. I think and I will argue for, all else being equal, the interest of a dog—let's make it simple, the pain and suffering of dogs—deserves the same concern as the equal amount of pain and suffering of human beings. I'm taking a dog as an example of something about which we all agree, but we could take any experiencing animal, like monkeys, dogs, and so on.

Wulfie's Pain, Bill's Pain—In a Lifeboat

If a dog is in a certain amount of pain (if he's bothered to a particular degree by a certain amount of pain) and a human being is bothered to the same degree—that is, is suffering as much as is the dog—they deserve equal concern. There is no reason to treat the human being more attentively or more seriously than the way we treat the dog. So here's the picture: so you have this guy. He's from Venus. He's a Venusian doctor/vet. I don't want to make him a human being and I don't want to make him a dog. So he's from Venus and he's equally good at being a human doctor and a dog doctor. He's perfectly skilled. And here's some human being that I know, let's call him Bill.

Here's Wulfie. Wulfie is any dog: little tail, smiling, happy dog. All right, so this dog is in a certain amount of pain. Bill also has a certain amount of pain. And let's say that they're in the same amount of pain. The human and the dog are suffering equally. The key point is suffering; how much the individual is being 'bothered by' the pain, frustration, depression, or whatever. They both have this Venusian, nonhuman/nondog vet/human doctor. They both run into the doctor's office at the very same time. So there's no "You were here first I have to treat you first," and again, no other complications. I don't want to hear that this guy's brother is the President of the United States and the world will end if you don't x or y—I want everything else held equal. So the claim is this: the doctor/vet has no reason to preferentially treat one rather than the other. Another way of saying this is that species membership, species identity, what species you are is, in and of itself, morally irrelevant. With me so far? Like interest(s) should be treated equally.

There's a common mistake in this line of reasoning I want to discuss briefly. Let's say now I've got a boat. There's the dog, and here's the human being and they're both drowning. They're flailing around in the water and they're going to both drown in a minute. And, of course, you can only save one. Now make the guy in the boat from Venus again because I don't want to get into technicalities about cospecies loyalties or species identity. So there's a human being drowning, a dog drowning and some guy who's not human or dog and he can only save one. Does my view entail that he should flip a coin? No! And you say, "How can that be? It looks like everything you just said suggests the Venusian should treat the interest of the human equally to the interest of the dog." While like interests should be considered equally, it may be that the human has a greater interest in survival than the dog. They're both going to die—this guy will die if you don't save him and the dog will die if you save the guy—but that doesn't mean they have an equal interest in survival.

I'm in the boat again, and now there's two human beings, drowning, and again I can save only one. Everything else is equal other than one guy is ninety-two years old. He's kind of sickly, he's done everything in his life, and he doesn't have that much interest in continuing it. The other guy is twenty-one. He's young, he's vibrant, he's got a lot of friends, a girlfriend, he's looking forward to next football game, and so on—you got the idea. If I save him, the old guy will die. That's true, but the two don't have equal interest in living. This young guy, by stipulation, has a lot more interest in survival than the old guy. So everything else being equal, you should save the young guy.

OK, so now let's go back to the first case. It may well be that the human being has a greater interest in surviving, in continuing to live, than the dog does. So I didn't say you treat everyone's interests equally. This really would be absurd; isn't it nutty, as a matter of course, to consider unlike interests similarly? All else being equal, the Venusian should save the human over the dog if the human has a greater interest in survival than does the dog, and he should save the dog over the human if the dog has a greater interest in survival than the human. Now philosophers argue whether nonhuman animals can have an interest in survival. It seems clear to me that they can, but this is another story. The point is that species identity doesn't, in itself, play a morally decisive role in these kinds of 'lifeboat' cases.

One reason for thinking that humans usually have greater interest in survival than dogs is that human beings can make plans for the future. I'm going to go to next week's party, I'm going to get married, I'm going to get divorced, I'm going to get remarried, I'm going to get redivorced . . . Or you're going to go to graduate school and you're going to have children, and you're going to become a doctor, you're going to go to Europe, and so forth. Now does the dog have any of those plans? Presumably not. The dog doesn't sit down and say, "Boy! Next week I'm going to have some kibble and then I'm going to go to PetSmart." The point here is, all else being equal, it looks like the human being, by virtue of the fact that he can make those plans about the future and project himself in the future, would have a greater interest in survival than the normal dog does. Not every human, I agree: the ninety-two-year-old might not give a damn about next week. But between the average twenty-year-old human and the four-year-old dog, you might argue the human has a greater interest in survival than the dog because, unlike the dog, the human has certain mental capacities—namely the ability to project or to think of himself or herself enduring in the future—that makes his interest in survival greater than the dog's. If that's the case, all else being equal, I should save the human. And maybe typically that's true. Maybe the typical human does have a greater interest in surviving than the typical dog.

Notice, though, that this result isn't speciesist. That is, it doesn't depend on the species membership of the individual. It depends on the specific mental capacities of the individual involved. Again, you can have some 140-year-old who doesn't give a damn about the next day. He's had it. If the dog, although he may not being able to think about how he'll be a month from now, at least has thoughts of how he'll be a few minutes from now, then I'd save the dog. So, if I should save the human, it isn't because he's a human that I should save him, but it's because he has a greater interest in survival. So this still conforms to my view that, in and of itself, species is irrelevant. The species membership or the species identity should never, in and of itself, be a factor when you morally deliberate about what to do. Now that I think this is a somewhat counterintuitive thesis. But I think it's correct.

Finally, if what we have argued is correct, there are significant implications for the use of animals in broader contexts, such as in biomedical research. If nonhuman animals share an equal moral status with humans—if, that is, their interests and lives are as morally significant as the interests and lives of humans—biomedical research should be greatly curtailed. The governing question for permissibility would be: Would we allow the same kind of research if the subjects of the research were human? Only an affirmative answer would morally permit the experimentation on nonhuman animals. Admittedly, there are nuances to this position, but the general conclusion to draw from our discussion is that invasive animal experimentation would not be allowed except in the most extreme conditions, where extreme conditions refer to circumstances in which we would deem it morally permissible to subject members of our own species to the same types of procedures.

References and Recommended Reading

Almeida, Michael, J., and Mark H. Bernstein. 2000. "Opportunistic Carnivorism." *Journal of Applied Philosophy* 17 (2): 205–11. doi:10.1111/1468-5930.00154.

Bernstein, Mark H. 1998. *On Moral Considerability: An Essay on Who Morally Matters.* Oxford: Oxford University Press.

Bernstein, Mark H. 2004. *Without a Tear: Our Tragic Relationship with Animals.* Chicago: University of Illinois Press.

Block, Ned. 1978. "Troubles with Functionalism." In *Perception and Cognition: Issues in the Foundations of Psychology*, edited by C. W. Savage, 261–326. Minneapolis: University of Minnesota Press.

Haidt, Jonathan. 2006. *The Happiness Hypothesis: Finding Modern Truth in Ancient Wisdom.* New York: Basic Books.

Henrich, Joseph, and Natalie Henrich. 2007. *Why Humans Cooperate: A Cultural and Evolutionary Perspective*. New York: Oxford University Press.

Kelly, Daniel. 2011. *Yuck! The Nature and Moral Significance of Disgust*. Cambridge: The MIT Press.

Kelly, Daniel, Kelby Mason, and Dennis Whitcomb. 2008. "Naturalization of Intentionality." In *Encyclopedia of Neuroscience*, edited by Marc D. Binder, Nobutaka Hirokawa, and Uwe Windhorst. New York: Springer.

Kelly, Daniel, and Stephen Stich. 2007. "Two Theories of the Cognitive Architecture Underlying Morality." In *The Innate Mind, Vol. 3: Foundations and Future Horizons*, edited by Peter Carruthers, Stephen Laurence, and Stephen Stich. New York: Oxford University Press.

McGinn, Colin. 1989. "Can We Solve the Mind-Body Problem?" *Mind* 98 (391): 349–66. doi:10.1093/mind/XCVIII.391.349.

Nado, Jennifer, Daniel Kelly, and Stephen Stich. 2009. "Moral Judgment." In *The Routledge Companion to the Philosophy of Psychology*, edited by John Symons and Paco Calvo. New York: Routledge.

Nagel, Thomas. 1974. "What is it Like to Be a Bat?" *Philosophical Review* 83 (4): 435–50. doi:10.2307/2183914.

Nordgreen, Janicke, Joseph P. Garner, Andrew Michael Janczak, Birgit Ranheim, William M. Muir, and Tor Einar Horsberg. 2009. "Thermonociception in Fish: Effects of Two Different Doses of Morphine on Thermal Threshold and Post-test Behaviour in Goldfish (Carassius auratus)." *Applied Animal Behaviour Science* 119 (1–2): 101–7. doi:10.1016/j.applanim.2009.03.015.

Richerson, Peter, and Richard Boyd. 2005. *Not by Genes Alone*. Chicago: University of Chicago Press.

Rothenberg, David. 2005. *Why Birds Sing: a Journey into the Mystery of Birdsong*. New York: Basic Books.

Simons, Ronald C. 1996. *Boo: Culture, Experience and the Startle Reflex*. New York: Oxford University Press.

Sperber, Dan. 1996. *Explaining Culture: A Naturalistic Approach*. New York: Blackwell Publishers.

Biographical Sketches

Daniel Kelly, http://web.ics.purdue.edu/~drkelly/, took his PhD from Rutgers University in 2009 and is an associate professor of Philosophy at Purdue. His research focuses on issues in philosophy of mind, cognitive science, and empirical moral psychology. He has written on issues of race and implicit bias, harm, and moral judgment. His recent book, *Yuck! The Nature and Moral Significance of Disgust* (2011), was published with MIT Press.

Mark Bernstein, http://www.cla.purdue.edu/philosophy/directory/?p=Mark_Bernstein, holds the Brewer Chair of Applied Ethics at Purdue University. Dr. Bernstein has written and worked on the problem of free will but finds his philosophical home in work on nonhuman animals and moral value. On that topic, he has written several books and articles including his 1998 *On Moral Considerability: An Essay About Who Morally Matters* and his 2004 *Without a Tear: Our Tragic Relationship with Animals.* Bernstein's latest book project explores the moral significance of friendship in human-animal relations and is currently under contract with Palgrave Macmillan.

Animal Pain: What It Is and Why It Matters (2011)

BERNARD ROLLIN

Control and Obligation: A Camel's Pain Matters Too

I'm going to tell you a famous story from antiquity that supports what I'm going to argue here. In this story, a man buys a camel from a wise man. And he says, "Tell me, oh savant, I've never had a camel before. Is there anything I need to know in particular in order to manage and husband the camel?" And the savant says, "Well, there's actually one thing. When springtime comes and mating season approaches they can get extremely hostile." (That's true of equids generally.) And the guy says, "Well, how do I manage that? What do I do?" And the wise man says, "Well, you must castrate the camel, and thereby diminish the urges which cause hostility and anger." The man asks, "How do I do that?" And the savant says, "It's extremely simple. You take a good-sized stone in each hand. You insert the testicles between the stones, and bring your hands smartly together." The guy says, "But, oh wise man, is that not extremely painful?" And the savant responds, "Not if you keep your thumbs out of the way." That's a metaphor for how many people have treated animal pain.

The basis of having a direct moral obligation to an entity is that what we do to that entity matters to it. We do not have moral obligations to rocks, wheelbarrows, tables, chairs, cars, diamonds, and other nonsentient entities. To be sure, we are morally obliged not to wreck tables, chairs, cars, diamonds; but

29

that is because what we do to them can matter to a human, not to them. It is wrong to wreck a table because it belongs to someone who is negatively impacted by its destruction. If one wrecks an unowned table, one has wronged no one, unless one person could conceivably have used the table.

Another way to state the same point is to affirm that only a sentient entity can have *intrinsic value*. That's one of those phrases that's thrown around a lot in ethics, like human dignity, which is one I really don't understand. The only sense I've been able to make of intrinsic value is that what happens to an entity matters to it, even if it does not matter to anything or anyone else. It cares about what happens to it and thus the valuing is intrinsic. That is not how all philosophers use the term, but it's the way I use it. Because that entity is capable of valuing what happens to it, either in a positive or negative way, such valuing is inherent in it. Rocks, tables, and hammers may have great instrumental or use value, but what happens to them does not matter to them. It is for this reason that one does not transgress against a table or a hammer when one destroys it or throws it away. Similarly, it is for the same reason we are held morally blameworthy when we use another human being simply as a tool. If you own the hammer it is perfectly permissible to throw it away when you are finished with it, but not so with the carpenter, even if one has hired him to do a job now completed. Any being capable of caring about what happens to it has intrinsic value by such caring, even if we in fact focus only on its instrumental value for us.

There are a whole slew of concepts related to this, like *end in itself* as opposed to *means*, which has been important for forty years in sexual ethics. The ability to experience pain is a sufficient condition, but not a necessary one, for a being to be morally considerable. Pain is an extremely valuable biological tool for survival. Though people may wish they did not feel pain when afflicted with it, a moment's reflection would reveal that those without that capability do not live a good life. People lacking the capability to feel pain, whether as a result of a well-characterized genetic malfunction, or as a result of a nerve-destroying disease like leprosy, have no alarm systems warning of injury or some other harm and eventually suffer shortened lifespan from disease or infection. Some people think, by the way, that leprosy is a bacterium that eats holes in your flesh. It isn't. Leprosy destroys your nerve endings and takes away your ability to feel burns or wounds, which, untreated, form skin sores. But the ability to feel pain is not a necessary condition for moral considerability. For example, a person or animal unable to feel pain warning of burns or infections resulting in loss of a limb would still be morally considerable and we would be blameworthy if we did not help a person or an animal preserve that limb, because after all being able to walk or run or to have two arms very much matters to the person or animal.

To take a more intriguing example, eighteenth-century British philosopher David Hume pointed out that organisms could have possibly evolved to be motivated to flee danger or injury, or to eat or to drink, not by pain, but by little pangs of pleasure that increase as one fills the relevant need, or escapes the harm (Norton 1984). Fair enough? I mean, you know, hunger could be delightful and when you fill that need it gets more delightful (same kind of motivational apparatus). In such a world mattering would be positive, not negative, but would still be based in sentience and awareness. In our world, however, the mattering necessary to survival is negative. Broadly put, injuries and unfulfilled needs ramify in pain. But physical pain is by no means the only morally relevant mattering. Fear, anxiety, loneliness, grief, and depression certainly do not equate to varieties of physical pain, but are certainly forms of mattering, right? I have a real problem with trying to list how much something like social isolation matters compared to physical pain. I don't think you can compare the two merely on one axis.

When I helped write the US laws for laboratory animals in the 1970s and then defended them before Congress in the 1980s, we included the need to control pain and distress and they passed successfully. Distress was essentially a placeholder for all these other levels of mattering. Pretty much to this day the USDA has had its hands full enforcing controls on pain, and so it never really got heavily into controls on distress. However, I will argue that some things that they would call distress are actually worse than physical pain. Considerably worse.

Telos as an Account of Mattering

Adequate morality for animals would include a full range of possible matterings unique to each kind of animal. In my account of animal ethics developed in a lot of places I have argued that the basis of our obligations to the animals under our aegis is the individual animals' nature that I call *telos*, following Aristotle in *De Anima* and *Metaphysics*. This is the unique set of traits and powers that make the animal what it is: the pigness of the pig, the dogness of the dog. This is well recognized of course in common sense, and exemplified in the song affirming, "fish gotta swim, birds gotta fly," right?—ideally said with a Brooklyn accent, of course. If, for example, we raise pigs in totally natural conditions, satisfying all aspects of pig nature from nest building to rooting to trading off care of young and all that, all of which are aborted in modern sow systems, I think we could say we understand happiness relative to that animal. When we fail to fully meet needs flowing from the *telos*, we harm the animal. While we do not have a word for the mattering implicit for a pig to forage or to build its nest, based on the way we keep them in modern confinement we can plainly see that each of these failures to meet what the animal is, by nature, is going to create a harm we are guilty of committing. The word "pain" simply does not capture the myriad ways our different treatments affect animals.

Sometimes not meeting other aspects of an individual's animal nature matters more to that animal than physical pain. The New Zealand ethologist Ron Kilgour showed in the seventies that cattle show more signs of stress when introduced into a herd of strange cows than when they are prodded with an electric prod (Kilgour 1978). That's interesting, isn't it? Chickens will go through electro shock grids in order to get access to the outdoors. Ethologists have given us myriad such examples. A cow, separated shortly after giving birth from its calf, the cow will bawl and moo for weeks, especially if it can see the calf. That's a particularly onerous thing to do. Separation does not cause physical pain, but it clearly causes suffering.

There is no simple word to express the many ways we can hurt animals besides creating physical pain. The ways are as countless as the multiplicity of needs flowing from variegated teloi and the interests that follow therefrom. So I am going to introduce a horrible expression used quite a bit in England now, since I couldn't come up with any better one. This is *negative mattering*, mattering in a negative way. Negative mattering means all actions or events that harm animals, from frightening an animal, to removing its young unnaturally early, to keeping it unable to move or socialize. Physical pain of course is the paradigm case of negative mattering, but covers only a small part of what the concept entails. And equally important, given the direction societal ethics is going, positive mattering is the opposite, right? All the things that satisfy the animals variegated needs emerge from its *telos*: freedom of movement, a sense of security, being with its offspring, and so on.

Not long ago when we passed legislation for lab animals, the scientific community complained that they didn't understand any of these things like pain and pain control, which was amusing because they lied for years saying they in fact controlled for these things. But all of a sudden they didn't get it. So I was asked to do a big book: fifty-five chapters, two volumes, on everything important in animal research that they needed to know about. I had a wonderful chapter from a guy named Hal Markowitz, who was a zoo ethologist. Markowitz told a story about a Portland, Oregon, zoo that created an enclosure for servals (Mellen et al. 1981). A serval is a kind of South African bobcat. So

Markowitz, sensing a good opportunity, had the zoo pay his way to the Kalahari where he watched servals in action. It turns out they spend 80% of their time predating low flying birds. They go down in these little bushes and twitch their little butts like a housecat, then a low flying bird comes by and they jump up and grab the bird. So he suggested that the feeding of horsemeat and chunks to the servals is not good—it's not what the servals need. The zoo people said, "What are we going to do? Throw them canaries?" That wouldn't fly very well. So he suggested that they grind up the rations, load them into an air powered meatball cannon that was computer driven, and randomly shoot them across the enclosure. Literally overnight the exhibit changed. It became the most popular exhibit in the zoo. Rather than servals laying around and languishing, they were engaged and active. This is a good example of actualizing the *telos* and making the animals happier.

I'm assuming that the level of activity and close attention the servals pay is indicative of happiness, but maybe not. But if our analysis from the book is correct, it is morally obligatory to expand the scope of veterinary medicine and animal welfare science to study all the ways that things can matter to animals negatively as well as positively, as society grows more concerned and more knowledgeable about animal treatment. In addition, it is necessary to attempt to understand which forms of negative mattering are the most problematic from an animal's perspective. We've already seen that being exposed to a new herd is more onerous for the cows than being electric shocked. The wildlife service in Colorado has confirmed the fact that coyotes will chew their legs off to escape a trap. This attests to the fact that immobilization is more aversive to them than physical pain, since it's obviously quite painful to chew your leg off.

The Study of Animal Pain

Studying these things without hurting the animal subjects is not a trivial challenge. The study of pain is similar. An excellent example of this arises from research into *learned helplessness*. Research in this area is actually so barbaric that it is banned in Britain and has been for years. It is, however, alive and well at the University of Colorado. Absurdly alleged to be a model for human depression, learned helplessness is achieved by subjecting an animal in a cage to an inescapable electric shock regardless of what it does. What eventually happens is the animal assumes the fetal position and does nothing. This horrible state almost never arises in nature and there is obviously no name for the horrible feeling engendered. Ironically now that we know about it we can assume that it occurs in animals housed in severely restrictive environments, such as sow stalls. That's learned helplessness.

The guru, the father of this field, is a man named Martin Seligman. I read a great article in the *American Psychologist* that basically argued that there are a lot of problems with Seligman, but the biggest problem with his work is that he uses dogs and college students, neither of whom are a model for human beings. Also necessary to study is the way in which the quality of negative experience changes with the cognitive states of animals. Mason's work on animal cognition and Jay Weiss's work on rhesus monkeys are examples. Mason showed in the seventies that animals' mental and cognitive states regarding cognitive experience absolutely modulate the degree to which an animal experiences the event as negative (Bernstein and Mason 1970; Mason 1971). Mason's work demonstrated that elevation of ambient temperature of mice to well above what's called the comfort zone, and even above the thermo neutral zone, varies in the degree to which the animal is disturbed. This variation is in accordance with whether the elevation is gradual, and thus can be cognitively processed by the animal as predictably rising, or sudden, where the animal has not had the ability and time to adjust its expectations to the ambient temperature or heat stress.

What's particularly amusing about Mason's paper is that he was living in the heyday of a period in which people denied animal consciousness. So when he writes that animals have cognitively processes and all this kind of thing, he makes sure to add a big footnote that says, "doubtless that will be explained soon physiologically" (1971). God forbid we say that animals think.

Similarly, Jay Weiss at Rockefeller showed that monkeys who are taught to anticipate and predict an electric shock—for example by ringing a bell prior to the shock—have a far less negative reaction to the stimulus than do those who do not know when the shock is coming (Weiss 1970; Weiss 1972). Basically this fits your own experience, right? These studies have profound implications for the nonpharmacological control of pain, which is actually a field in laboratory animal medicine. It has long been known that laboratory animals subjected to an invasive procedure followed by a reward have a less negative reaction, pain or otherwise, than those who are simply restrained. For this reason smart researchers train animals to the procedure. This is another great story: one of my friends was doing a vaccine shedding study and she went into the facility every day, played with each dog, drew blood and then gave them a cookie. One day, one of the dogs set up this incredible howl, the "kai yai yai yai yai" which is indicative of misery. She thought it had its foot caught in the cage door. So she went back to check the dog, and there was absolutely nothing wrong with him.

She started to leave again, and again the "kai yai yai yai yai." And she looked at her list and she'd forgotten to draw blood from him. He had not gotten his blood drawn and therefore had not gotten his treat, and then not gotten his playtime with her. It's a cute story and an important lesson.

The Australians in the seventies had a research colony of baboons trained so that they could blow a whistle and each baboon would present its arm for blood draw in return for a treat (personal experience). When I was at UC Davis in the early eighties they had their baboons in individual boxes or cages, and I said, "Why in the hell do you keep baboons like that? It violates their nature." And they said, "They're mean." I said, "Did it ever occur to you that you've got it ass backwards? They are mean because of how you keep them, not that you keep them that way for being mean." The Australians had demonstrated this.

Positive and Negative Mattering

It follows from what we've argued that all aspects of negative mattering should broadly enter in the field of pain control. Nonetheless there's a famous remark that physical pain is the worst of evils. At first blush this seems wrong. We are inclined to believe that the death is the worst of evils: common sense. But careful reflection reveals this is false. Even among humans, people will readily choose death over prolonged or intense physical pain, and even over emotional pain. This is evident in the worldwide thrust for assisted suicide on the part of those suffering intractable physical pain or mental anguish. Indeed, many people choose death not only over pain, but over helplessness, total dependence on others, loss of dignity through things like incontinence, and so forth. While there are many cases such as radical cancer treatment where people undergo prolonged and horrible pain in order to live, there also very many where death is chosen to forestall suffering on the part of human beings.

Since our topic is animal pain we must question how animals value death as compared to pain. This is an increasingly important conceptual-ethical issue in a world where ever increasing amounts of highly invasive therapy are being exported from human medicine to animal medicine, resulting in great amounts of suffering on the part of animal subjects. For example, this is particularly true in oncology where cost is no obstacle to moving human modalities to sick animals. Believe it or not, as early as 1982 the Wall Street Journal reported that at CSU clients were spending well over a hundred thousand dollars for treating their pets in heroic ways for cancer. However, from the animal's perspective, is the chance for extra life worth the significant extra suffering? This in turn leads to an interesting, ancillary question which was discussed at a recent British meeting on veterinary ethics: "Can an animal value life, per se, rather than its content?" (International Conference 2011).

To answer this question we must consider some conceptual differences between animal and human cognition. Human cognition is such that it can value long-term future goals and endure short-term negative experiences for the sake of achieving them. You can think of many examples. We undergo voluntary food restriction to lose weight or to look good in a bathing suit, and undergo the unpleasant experiences attendant in its wake. We memorize volumes of boring material to get admission into veterinary or human medical school. We endure the excruciating pain of cosmetic surgery to look better. We similarly endure chemotherapy, radiation, dialysis, physical therapy, and transplant surgeries to achieve a longer, better quality of life than we would without it, or in some cases merely to prolong life in order to see our children graduate, complete some opus, or see Ireland again.

The Concept of Death for Nonhuman Animals

In the case of animals there is no evidence, either empirical or conceptual, that they have the capability to weigh future benefits or possibilities against current misery. To entertain the belief that my current pain and distress resulting from the nausea of chemotherapy or some highly invasive surgery will be offset by the possibility of an indefinite amount of future time is taken to be axiomatic of human thinking. But such thinking requires complex cognitive machinery. For example, one needs temporal and abstract concepts such as possible future times and the ability to comparatively weigh them. Although I despise Heidegger, he defined very well the concept of Death as grasping the possibility of the impossibility of your being (Heidegger 1996). That's a pretty good definition for thinking about death, the ability to articulate possible suffering and so on. This is turn requires the possibly to think in a 'if then' hypothetical and counterfactual mode; that is, if I do not do x then y will occur. This mode of thinking seems to necessitate or require the ability to process symbols and combine them according to rules of syntax.

Fair enough so far? I have argued vigorously elsewhere against the Cartesian idea that animals lack thought and are simply robotic machines (Rollin 1989). I strongly believe that animals enjoy a rich mental life. It is also clear that animals have some concept of enduring objects, causality (against Kant, who denies this), and a limited concept of future possibilities (probably learned by association, else the dog would not expect to get fed, the cat would not await the mouse outside of its mouse hole, and the lion could not make the trajectory of its run such that it can intercept the gazelle.) And of course Darwin himself argued that animals display a full range of emotion (Darwin 1969).

Darwin did some very fascinating things with animals, including, by the way, arguing that earthworms have problem solving ability, in a book called *The Formation of Vegetable Moulds* (Darwin 1881). Just wait for the movie. But it is equally evident that an animal cannot weigh being treated for cancer against the suffering the treatment entails, cannot affirm or desire or even conceive of a desire to endure current suffering for the sake of future life, cannot understand that current suffering may be counterbalanced by future life, and cannot choose to lose a limb to preclude metastases of cancer. To treat animals morally and with respect, we need to consider their mentational limits. Paramount in importance is the extreme unlikelihood that they can understand the concepts of life and death in themselves rather than the pain and pleasure associated with extended life or death. To the animal mind, in a real sense there is only quality of life. That is, whether the experiential content is pleasant or unpleasant in all of the modes it is capable of, experiencing, whether that's boredom, stimulation, fear, loneliness, pain, hunger, or thirst. We have no reason to believe that an animal can grasp the notion of extended life, let alone make the choice of trading current suffering for it.

This in turn entails that we realistically assess what they're experiencing. For me, an animal *is* its pain. Pain is the animal, for it is incapable of anticipating or hoping for cessation of that pain. Thus when we are confronted with life-threatening illnesses that afflict our animals, it is not axiomatic that they should be treated at whatever qualitative experiential cost that may entail. The owner may consider the suffering a treatment modality entails a small price to pay for extra life, but the animal neither values nor comprehends the extra life, let alone the tradeoff.

A very important corollary emerges from our discussion. Our argument says: animals have no concept of death or life, and consequently cannot value life or death more than pain. We have also indicated that people sometimes value death over pain as a way of ending pain. If this is true of humans, it would be true of animals who cannot value life at all. Thus in a sense pain may well be worse for animals than for humans as they cannot rationalize its acceptance by appeal to future life without pain. As I've said in other writings, a traditional argument affirms that human pain is worse than animal pain because humans can anticipate and fear pain very imaginatively before it happens (Rollin 1989). When we plan to visit the dentist, we worry that the dentist is Joseph Mengele because we saw *Marathon Man* or we might worry that the dentist had a fight with his wife. That is why dentists have such a high rate of suicide—they're confronted with people who are terrified of them but fawningly smile.

Aside from the fact that animals too can fear imminent pain, evidenced, for example, when they cringe before threatening upraised hands, the same logic dictates that animals cannot look forward to a time without the pain. Their entire universe is their pain—they cannot have any hope. That thought hit about twenty years ago and just knocked me on my butt. Can you imagine life without hope?

An obvious question coming from scientists, but not typically from ordinary people, arises here. How can we know that animals experience all or any of the negative states we have enumerated? Are we not guilty of the dreaded career-ending sin of anthropomorphism? How do we know that animals that appear to us have a full range of mental states, particularly pain and suffering, are not simply cuckoo clocks as Descartes argued? Such agnosticism about animal feeling is not simply abstract speculation. It was, in fact as some of you may know, assumed by Descartes' biologist followers as a justification for doing horribly invasive manipulations, including genuine vivisection. Vivisection today is used to mean any animal experimentation. What it meant then was cutting the animal alive. And there are actually drawings from people who visited the Port Royal school of Cartesians where the animals are chained to tables and walls and cut on while they are alive. Of course the 'saintly' father Malebranche, the Cartesian, used to reassure students that Descartes has freed us from the anxiety that the dog moaning in pain is just a cuckoo clock. And the same saintly father Malebranche would bring his mastiff bitch to class, pregnant, and kick her in the stomach, to help the student grow sensitized to the idea that animals don't feel pain; it's just a mechanistic reflex.

Responses to Scientific Ideology

The notion that we need to be agnostic or downright atheistic about animal minds including pain because we cannot verify them through our experience was pervasive through European science and philosophy. In the US it became a mainstay of what I've called in another book *scientific ideology*, (Rollin 2011, 91; 2006b) the uncriticized dogma taught to young scientists through most of the twentieth century, despite its patent ignoring of Darwinian phylogenetic continuity. Together with the equally pernicious notion that science has nothing to do with values and particularly ethics, this provided the complete justification for hurting animals in science without providing any pain control. And, while they were at it, these dogmatic scientists didn't allow for mind in human babies either because they bought another Cartesian argument claiming that you needed to have language to feel pain. I've done over sixteen hundred talks and the worst response I ever got was when I went to the Pain Society, the society for the study of pain, and debated with the MDs. They were so angry that they just walked out when I was done. They were white with rage, because I dared to affirm that babies feel pain. One guy tried to respond to me and asked if I were circumcised. I said "As a matter of fact I was, not that it's any of your damn business, but why?" He said, "You remember it?" I said "no." He said, "So what difference does it make?" You have to answer that claim. How should you respond?

Well number one, whether I remember that experience now or not, I felt it at the time. That is itself morally questionable for innocents. Even worse than that, we know for a fact that babies that experience uncontrolled pain have a repatterning of the nervous system that make them susceptible in adulthood to chronic pain. You see?

That felt pain was denied or ignored for much of the twentieth century is easy to evidence in an objective way. I wrote and advocated for federal laws protecting laboratory animals and got them passed in 1985. I repeatedly came up against the denial of pain by the scientific community in both objectively documentable ways and in personal experiences, some of which I'll document for you because they are interesting stories. One of my favorites took place when I went to England in 1982 as an UFAW Lecturer, (that's a big fancy animal welfare organization that all the intellectuals belong to). I was asked to comment on a paper by a very famous pain physiologist. I had fifteen minutes to respond, to take him apart, after he had argued for an hour that the electrochemical activity in the cerebral cortex of a dog was different from that of people in a marked way, and, since that is the information processing area, the dog didn't really feel pain in any way we can identify with. I mean, I'm long winded as you're now learning, but this was the shortest response I've ever given. I said, "Dr. X, you're a very famous pain physiologist; you even have cells named after you. If I'm not mistaken, sir, your entire career has been pain research on dogs." He said yes. And I said, "And you extrapolate the results to people right?" He said, "Of course. That's why I do it." I said, "Good. Either your speech is false or your life's work is." You know? That was fun.

You will look in vain at veterinary textbooks published before the late nineteen eighties to find any mention of the fact that pain hurts, that pain is felt. Do you know the different between anesthesia and analgesia? Anesthesia means, literally, "no feeling." So when you get an operation, an appendectomy for instance, they give you a general anesthetic so you don't feel anywhere, though it's more complicated than that. If you go to the dentist, you get a local anesthetic so you don't feel anything in your jaw. Analgesia, on the other hand, raises pain tolerance. One of the things that I discovered in the seventies was that the research community did not use analgesics on animals. For example, when they studied fracture fixation and broke legs, they didn't give the animals anything for the pain. I knew this anecdotally from technicians in medical schools and from some decent researchers. So I went before Congress with a bill to mandate the control of pain. Congress said, "Well to be honest with you, the research community gives us a lot of money for our war chests for elections. You don't give us anything. They don't want this. So if you want it you better give us damn good proof that they're not doing when they say they are." Well that was pretty tough, a challenge that bothered me for a week. How the hell was I going to prove that they were not controlling pain? I could bring in technicians who told me, but they'd send in a Dean who'd say, "He doesn't understand." But it finally dawned on me. I went to the Library of Congress. We didn't have PCs that were very powerful then, but the Library of Congress had search tools that could do a huge literature search. I asked a friend to search for laboratory animal analgesia. Unsurprisingly, he found zero papers. We expanded it to animal analgesia and we found two, one of which said there should be papers. The other said we don't know anything.

When I presented that evidence to Congress, they basically agreed to consider the law, and eventually it passed. I did a literature search just recently. Do you know how many papers there are on analgesia now? As of that search, there were 11,866 papers—and hopefully a correlative increase in the use of analgesics on laboratory animals. I lectured in Italy in May 2011 and I told some of this to the audience because I had a new book coming out in Italian translation. As a joke I said, "Maybe this will keep me out of Hell," you know, if God really sees every sparrow fall and all that. I got forty emails from Italians in my audiences saying, "Don't worry we're absolutely sure you're not going to Hell." Well, can I take succor from that? I'm not sure, but it was a nice gesture.

Current Issues in Animal Pain

So where is the issue in pain control in animals now? I mean, if research animals are having their pain controlled, then which animals are not? The answer is agricultural animals. For instance, how do the cattle people, the best of agriculturists because they still practice husbandry, produce permanent identification? They use hot iron branding, inflicting a third-degree burn on the animal, destroying the melanocytes, and creating a permanent scar that is indicative of ownership. My colleague and I in veterinary medicine did the first and only study to date on the use of a local anaesthetic prior to branding. Lo and behold, it hurts less if you have a local anaesthetic on board, even if it's washed out after a while. Hopefully we can continue to make some more progress on that.

Ordinary common sense throughout the history of philosophy has been in contradistinction to what I call scientific ideology, or scientific common sense, even before Descartes ever denied that animals felt pain. Much more pervasive is the notion that animals, despite the fact that they feel pain, have very little moral status. The Bible, to its credit, and the Old Testament primarily, was patently clear not only about the fact of animal pain, but also about its moral relevance. The Pentateuch is clear about forbidding practices that cause physical pain, but also other notions of negative mattering, and about the degree to which bad animal treatment can erode human sensitivity. Such passages as the extension of Sabbath rest period to animals, or forbidding the yoking of oxen and asses to the same plow for reasons of the significant strength differential, display great sensitivity to negative mattering other than physical pain. There is actually a passage that says if you take eggs out of a bird's nest, leave some for the bird to avoid distress. Kosher slaughter, involving someone who is anatomically knowledgeable cutting the jugular vein with a razor sharp knife in a calm environment, was intended in antiquity as an alternative to bludgeoning, with big sledgehammer type things, and was indeed a quick route to unconsciousness. Ironically, when teamed up with the time pressure of modern slaughterhouses, kosher slaughter becomes grossly inhumane.

The Rabbis, to their credit, forbid enjoying any pleasure from animal suffering. So Jews don't hunt unless they hunt for survival. To his credit, a prophet of the Mormon Church from twenty years ago, Spencer Kimball, had a specific passage in which he forbade Mormons from hunting for pleasure, which was pretty good since if you've been to Utah you know it's a major hunting, shooting, survivalist culture. So he displayed some courage in saying that. The Hebrew concept of suffering of living things as I mentioned is called *tsaar baalay chaim*. In the Christian tradition cruelty to animals is similarly forbidden despite the fact that animals, lacking souls, have no moral status. But St. Thomas Aquinas said that animal cruelty is nonetheless forbidden on the psychological principle that animals act like people and that those who abuse animals will end up abusing humans: an impossibility were not the suffering very, very similar. And in fact the last fifteen convicted serial killers, all the kids who shot up their classmates, and eighty percent of the violent offenders in Leavenworth have histories of early animal abuse. So the idea is to cut that off at its root. Who's guilty of animal cruelty, statistically? Men or women? Old men or young men? The answer is very young men: men in the early teenage years, when they're bags of hormones. Most of those kids grow out of it. But the other group, of course, is made of psychopaths who begin with animals and graduate to people (Knoll 2006).

If common sense readily acknowledges pain and other forms of mattering, how did science come to be agnostic about animal pain? First of all, modern science as opposed to Aristotelian science, feels no reason to accord with common sense, or ordinary experience. In fact it becomes a badge of modern science that it supersedes common sense. Take, for example, Einsteinian physics of velocities not being additive such that nothing can go faster than the speed of light. If I'm riding my motorcycle and shining my light toward you it's going at 186,000 miles per second squared (miles/s^2). If I'm riding my bike at 100 miles/s^2 toward you, the light is not going 186,000 plus 100 miles/s^2: it is still just going 186,000. And that is a radical violation of common sense. Of course you see this scientific hubris in Descartes, the father of modern science, when in his first meditation he does his best to discredit common sense and what you know through experience (Descartes 1984, 12–16).

As I said, scholars pretty much all agree that no one denied feeling to animals before Descartes. What did occur historically was the failure to reckon the full impact of pain and suffering on the moral status of animals. In other words, animal pain was acknowledged to be real but not credited with a great deal of ethical importance. There are several reasons for this. Egalitarian thought was historically quite rare before the enlightenment, and even though all humans were seen as feeling pain or being subjects of positive and negative mattering, some people's mattering was less important than others. It is often said if animals could speak we would consider their pain significant, yet there are indelible and perfectly clear records in the nineteenth century when slaves were subjected to painful medical experiments. They could speak, but nobody gave a damn about their pain. Concern about pain has a lot to do with sociocultural status. There is a wonderful book that you should read if you're interested in this called the *Calculus of Suffering* by a historian named Martin Pernick (Pernick 1985). Who got the most control of pain in the nineteenth century? Women. Rich women, except for pain of childbirth, which they deserved because of Eve's transgression. Who among men got the least pain control? Soldiers, sailors, construction workers—because they were tough. Children got a lot of pain control because they were innocent. Today it's just the opposite: child pain is much less controlled than adult pain. Why? Because kids can't sue.

Furthermore, the main use of animals historically has been in agriculture; food, fiber, locomotion, and power. The traditional key to success in agriculture until the twentieth century industrialization was applying good husbandry to the animals you raised. Husbandry means putting animals in the optimal environment they are biologically suited for and improving their ability to survive and thrive by provision of food during famine, water doing drought, medical attention, help in birthing, attention during disasters, protection from predation. In fact, this is so powerful an icon in western mind, that when the psalmists searched for a metaphor for God's relationship to humans, what did they use? Recall the twenty-third psalm: "the Lord is my shepherd." And as I wrote in an article I published in the Christian Century (not bad for a Jewish atheist), "It does not say, 'the Lord is the manager of premium standard farms. He cometh to me in coveralls. He maketh me lie on slats in small stalls in my own excrement for the duration of my productive life'" (Rollin 2001). They gave me an argument about that.

Good husbandry was essential to agricultural success; as such you really didn't need animal ethics because self-interest was a much more powerful motivator than ethics. If you hurt the animals you hurt your own self-interest. It's still known to this day that if you raise cows in a nonindustrial way, the number one variable correlated with milk production is the personality of the herd person. Now husbandry is less important when we have artificial hormones like bST.[1] One way to characterize the rise of massive concern for animal treatment in the twentieth century was the social realization in the sixties that industry had replaced husbandry. Whereas historically you had to treat the animal well to make profit—in other words put square pegs in square holes and round pegs in round holes—now we have technological sanders that can force square pegs into round holes. So the sow lives and reproduces, but she's not happy. Well, who cares? The stockholders are making money. And in the case of those people who are not motivated by self-interest, the sadists and psychopaths, we have the anticruelty laws.[2]

What percentage, would you say, of animal suffering in human hands is a result of what these cruelty laws cover; that is, the deliberate, sadistic, purposeless, deviant, senseless infliction of pain and suffering on an animal? Tying a firecracker to a cat's tail, burying a cat up to its neck and running over it with a lawn mower, beating the dog to death with baseball bats: how much of animal suffering is a result of that? Thankfully, not much. In fact every audience that I've ever spoken to on this has said less than one percent. What proof of that? We produce eleven billion chickens for food last year. Ten percent of those were bruised or fractured by the time they go to market.

Ten percent of eleven billion is one point one billion. If there were one point one billion instances of cruelty you couldn't step out of your house without seeing someone setting fire to a horse and running over a cat and so forth and so on. Thankfully overt cruelty is very minimal. Although have you seen the video of the beating of bull calves to death with hammers on that Texas calf ranch? Look it up. It'll change you forever. I'm proud to say that the Colorado Livestock Association said that guy should be horsewhipped. They had to cut that and say that he should be "prosecuted to the fullest extent of the law," but what they meant was he should be horsewhipped.

The Denial of Pain in Nonhumans

So as a final point, where does the denial of pain come from if it's inimical to common sense? I told you it comes from Descartes and from scientific ideology, but why and how? The greatest skeptic in western philosophy was David Hume (Norton 1984). Hume denied mind, body, causality, a unified self, mathematical physics, uniformity of nature, and God while he was at it. Yet, the same Hume warns against applying this to life. He says be a philosopher but, in the end, be a person, and see the world a certain sort of way. The same Hume, who denied so many things, takes the people who deny animal thinking or animal feeling to task. He said it is so obvious that it escapes only the most stupid—meaning Descartes.

Hume begins with sense experience, deduces therefrom the paradoxicality of metaphysical speculation, and returns to common sense (Norton 1984). Descartes on the other hand never feels compelled to return to ordinary experience after rejecting it. Because, according to Descartes, a false sense experience is qualitatively identical to a true one, you see? False experiences don't wear a convention badge announcing, "Hi, I'm a false experience." Otherwise, we could never be fooled by the senses. Therefore we cannot make a soundly based distinction between illusion and reality and therefore reject the senses as the sound source of knowledge. My colleague and mentor Professor Arthur Danto has pointed out that this amounts to saying that since sometimes the senses can be wrong we should always reject them, which is as crazy as saying that since sometimes the senses can be right we should always accept them

(Danto 1968). Descartes has essentially taken a practical problem of ordinary life and turned it into an insoluble conundrum. In ordinary life we're indeed fooled by our senses when we see double, or when we see a wet spot on the pavement on a hot day, or when that wonderful line of W. B. Yeats in the poem *The Apparitions*, "fifteen apparitions have I seen, the worst a coat upon a coat hanger" (Yeats 1997). Do you know what he meant by that? You come home late at night, you're tired, maybe you're drunk, and you left the raincoat on the coat hanger and it's picked up by the moonlight and AHHHH! An intruder! Right? These things happen.

We correct these misperceptions by more systematic perceptions. We go to the coat rack, we see it's not a bogeyman, and we reject it. To be sure, as Descartes would point out, this is in some way circular: we check perception by using perception. But it works and there is no real reason to believe there is a wet spot on the highway, or an intruder, or two copies of my house when I squeeze on my eyeball. Only a madman worries about what is not real being real, and what is real being not real. I believe, and we can argue about this, we perceive pain and emotion in others. We do not infer it or construct it, as the positivists would have us believe. A child watching a dog writhe or listening it to it yelp after a bee sting immediately understands the presence of pain in the animal. Similarly, with another child who falls on his or her face and starts to bleed. The only skepticism you can generate in such situations is if we have evidence that the pained individual is faking: the child is in a play and is rehearsing, or has told his friend, "Watch me upset my mother." Indeed after we watch a great actor simulating grief and pain and know that such faking is possible, we still believe behavior in a normal context betokens pain. And in fact the Greeks knew that for a great actor who really simulates pain, he often feels it and other emotions. Skepticism under such circumstances is forced and unnatural. If you cut your leg while chopping wood and are screaming hysterically, it is absurd to ask "hmm, how do we know he isn't faking?" in British analytic fashion.

The same holds true of animals. Descartes's denial of pain is based on the faulty logic seen above. Of course it was really convenient because it allowed people to cut up animals while they were alive without feeling guilty. I published a book in 1989 which was pretty influential in restoring the control of pain in the English speaking world. When I was researching that book and I found that veterinarians had nothing in textbooks about pain control, which I found extraordinary, I called up a veterinarian who had written about pain control. He was an Italian guy from the Bronx. I asked him my question; I said,

"Do you ever encounter people who deny that animals feel pain?" "Of course," he said. I said, "How do you respond to them?" "You want to know?," he replied, "I'll tell you how I respond to them. I tell them to run a little experiment. Take an intact male Rottweiler, put him on your examining table, carefully fit a vise grip to his nuts, and then squeeze, and he'll show you he's in pain." And I said, "How?" "He'll rip your goddamned face off." OK, good point.

This is an appropriate response to a genuine query and should end the conversation. If the questioner remains skeptical about pain in animals, we assume there is something wrong with them. Just like a little kid who keeps saying why, "Why is the sky blue," "Well this refraction . . ." "But why? But why? But why? But why?" At a certain point you just turn him off. This is how we judge human intentions in ordinary life or in courtrooms. Skepticism about pain or other mental states in humans or animals has perfectly legitimate resolutions in ordinary life. For purposes of ordinary life it is legitimate to reject unending skepticism. If someone is truly skeptical about animal pain, it is appropriate to refer them to congruence of the physiological substratum in humans and animals, to the fact that we study pain in animals and extrapolate to people, to the fact that anesthetics and analgesics work on animals, to the similarities in pain behavior, to our natural empathy for a creature in pain, and to the fact that such skepticism must inexorably extend to humans as well.

But animals do fake pain, just like people do. Did you know that? I experienced a wonderful case of that when I had my Great Dane in New York and she developed a really major limp. I went to the animal medical center and they said, "Oh my god, she has osteochondritis dessicans." I said, "Oh my god, we have to do surgery." So I said, "I'm going to take her (Helga was her name) down to the park, Riverside Park, for the last time, to see her friends." So she limps down laboriously to the park and I'm in tears. We get to the park and I remove her leash. She takes off like an arrow after her friends and comes running back. I said, "Helga, you're in trouble for that, you know? You're lying to me about the pain." The limp didn't come back until two years later. We went to northern Canada in the woods and we let her run without a leash. And she didn't come back. She took off after some creature and we're screaming "Helga, Helga!" in the middle of the North Woods. Fifteen minutes later she comes back. And I said, "Oh, Helga, you've had it." And she limped.

There are actually scholarly articles about this. Maybe scientists are so skeptical as to say, "Well, maybe they show all the evidence of pain, but we still want to be skeptical about whether they feel it." If science is that skeptical about empirical evidence, they have to be that skeptical about the reports of data from other scientists. For instance, how do we know they perceive as we do? They take private subjective experiences and construct the intersubjective world out of it. If you can't do that you can't do science. And if you can do science, you can attribute pain to animals.

Bertrand Russell pointed out that we have really no way of rejecting the hypothesis, scientifically, that the world was created three seconds ago with all the fossils, us with all our memories (Russell 1959). So, why do science? You can't be selective about that kind of skepticism. It's too powerful. So the practice of science belies its ideological presuppositions. And of course the extrapolation of pain research to people also belies that.

Once I was invited to AALAS, a lab animal science meeting. It's technicians, PhDs, and DVMs, mostly. There are about five to ten thousand people at these meetings. They asked me to defend the idea that we need a law mandating pain control, and of course I got up and did it. One of the veterinarians said, "Well why should we do that? We don't even know for sure that animals feel pain." I knew this guy pretty well and I didn't say anything. Five years later the law passed and I called him up and said, "Well, you were skeptical about animal pain. Presumably now you can't be. You have to control the pain. What are you going to do?" And he said "Oh, I use this, this, this, this, or this." I said, "Wow, what happened? You were skeptical that they even feel pain five years ago." And he says, "Oh, all human pain killers are tested on animals and have been for seventy five years." You know what that shows you? It shows you that your metaphysic can so color your gestalt that it overrides common sense. At the point that it was mandated that animals feel pain, and he had to control it, all of a sudden he began to see all that data as relevant to animal pain.

In conclusion, in the end there are no sound reasons for rejecting knowledge of animal pain and of other forms of negative mattering in animals. Once that hurdle is cleared, science must work assiduously to classify, understand, and mitigate all instances of negative mattering occasioned in animals by human use, as well as to understand and maximize all modes of positive mattering. As society develops more and more fine grained concern about animal pain, you are going to see more and more emphasis on what makes animals happy, not only on stopping what makes them miserable.

Notes

1. Bovine somatotropin, or bST, is a growth hormone developed to aid in milk production of dairy cattle.
2. Laboratory Animal Welfare Act of 1966, P.L. 89-544.

References and Recommended Reading

Aristotle. 1896. *Metaphysics*. Translated by J. H. MacMahon. London: G. Bell and Sons.

Aristotle. 1991. *De Anima*. Translated by R. D. Hicks. Amherst, NY: Prometheus Books.

Bernstein, Stephen, and William A. Mason. 1970. "Effects of Age and Stimulus Conditions on the Emotional Responses of Rhesus Monkeys: Differential Responses to Frustration and to Fear Stimuli." *Developmental Psychobiology* 3 (1): 5–12. doi:10.1002/dev.420030104.

Danto, Arthur. 1968. *Analytical Philosophy of Knowledge*. Cambridge: Cambridge University Press.

Darwin, Charles. 1969. *The Expression of Emotion in Man and Animals*. Westport, CT: Greenwood Press.

Darwin, Charles. 1881. *The Formation of Vegetable Mould through the Action of Worms, with Observations on their Habits*. London: John Murray. doi:10.5962/bhl. title.48549.

Descartes, René. 1984. "Meditations on First Philosophy." In *The Philosophical Writings of Descartes: Volume II*. Translated by John Cottingham, Robert Stoothoff, and Dugald Murdoch, 1–50. Cambridge: Cambridge University Press.

Heidegger, Martin. 1996. *Being and Time*. Translated by Joan Stambaugh. Albany: SUNY Press.

International Conference on Veterinary and Animal Ethics. 2011. ICVAE Programme. *Veterinary Ethics*. http://www.vetethics.com/events/icvae1/programme.

Kilgour, Ronald. 1978. "The Application of Animal Behavior and the Humane Care of Farm Animals." *Journal of Animal Science* 46 (5): 1478–86.

Knoll, James. 2006. "Serial Murder: A Forensic Psychiatric Perspective." *Psychiatric Times* (March): 64–68.

Mason, John W. 1971. "A Re-evaluation of the Concept of 'Non-specificity' in Stress Theory." *Journal of Psychiatric Research* 8 (3–4): 323–33. doi:10.1016/0022-3956(71)90028-8.

Mellen, Jill D., Victor J. Stevens, and Hal Markowitz. 1981. "Environmental Enrichment for Servals, Indian Elephants, and Canadian Otters at Washington Park Zoo, Portland." *International Zoo Yearbook* 21 (1): 196–201. doi:10.1111/j.1748-1090.1981. tb01981.x.

Norton, David. 1984. *David Hume: Common Sense Moralist, Skeptical Metaphysician*. Princeton: Princeton University Press.

Pernick, Martin S. 1985. *A Calculus of Suffering: Pain, Professionalism, and Anesthesia in Nineteenth-Century America*. New York: Columbia University Press.

Preece, Rod. 2007. "Thoughts Out of Season on the History of Animal Ethics." *Society & Animals* 15 (4): 365–78. doi:10.1163/156853007X235537.

Preece, Rod, and David Fraser. 2000. "The Status of Animals in Biblical and Christian Thought: A Study in Colliding Values." *Society & Animals* 8 (3): 245–63. doi:10.1163/156853000511113.

Rollin, Bernard. 1989. *The Unheeded Cry: Animal Consciousness, Animal Pain and Science*. Cambridge: Cambridge University Press.

Rollin, Bernard. 1995. *Farm Animal Welfare: Social, Bioethical, and Research Issues.* Ames, IA: Iowa State University Press.

Rollin, Bernard. 2001. "Farm Factories: The End of Animal Husbandry." *The Christian Century* 118 (5): 26–29.

Rollin, Bernard. 2006a. *Animal Rights and Human Morality.* 3rd ed. Buffalo, NY: Prometheus Books.

Rollin, Bernard. 2006b. *Science and Ethics.* New York: Cambridge University Press. doi:10.1017/CBO9780511617218.

Rollin, Bernard. 2006c. "Euthanasia and Quality of Life." *Journal of the American Veterinary Medicine Association* 228 (7): 1014–16. doi:10.2460/javma.228.7.1014.

Rollin, Bernard. 2011. *Putting the Horse Before Descartes: My Life's Work with Animals.* Philadelphia: Temple University Press.

Russell, Bernard. 1959. *The Problems of Philosophy.* Oxford: Oxford University Press.

Weiss, Jay. 1970. "Somatic Effects of Predictable and Unpredictable Shock." *Psychosomatic Medicine* 32 (4): 397–408.

Weiss, Jay. 1972. "Psychological Factors in Stress and Disease." *Scientific American* 226 (6): 104–13. doi:10.1038/scientificamerican0672-104.

Yeats, William B. 1997. "The Apparitions." *Last Poems.* Ithaca, NY: Cornell University Press.

Biographical Sketch

Bernard Rollin, http://central.colostate.edu/people/brollin/, is University Distinguished Professor, professor of philosophy, professor of animal sciences, professor of biomedical sciences, and University Bioethicist at Colorado State University. Rollin received his PhD from Columbia University in 1972 and dedicated his career to advocating for the ethical treatment of nonhuman animals. To this end, Rollin has testified before Congress, consulted internationally, and done pioneering work in academia. Rollin has lectured more than 1600 times and published over 300 papers and at least 15 books.

ENVIRONMENT

Ethical thinking about the natural environment developed only recently as a direct result of continued moral expansionism. This expansion of moral thinking out into the natural world was driven by the ethical considerations first of the human rights movement and then of the animal rights movement, but it was equally the result of developments in our scientific thinking about ecology, the study of relationships within the natural world. Ecological research in the US began to pick up steam as far back as the early 1930s. It was the economic thinking of that era (thinking that continues to pervade environmental policy even today) that drove early discussions about human obligations to the natural world. As production and "progress" began to threaten the health of ecosystems, the value of those natural structures was found not only in their monetary worth, but in something deeper—something inherent in the structures themselves.

Only by distinguishing between inherent and mere use value can an account of moral worth in the natural world give us a foundation from which to develop robust policy around environmental conservation that supports biodiversity and environmental health. Such an account of *inherent worth* might be supplemented by accounts of *use value*, the value a thing has for another, but the former supports a ordering of priorities that can overcome economic concerns in a way that a solely instrumental value cannot. Thus, the essays in this section are framed as an exploration of what we consider an *ecological ethic*, that is, a coherent theory of environmental value grounded in science that prioritizes individual life and the natural world in terms of its interrelations. This perspective is an important aspect of our broader conception of bioethics.

The global situation of the twenty-first century is unique. Global policy (not to mention global ethics) is only beginning to come to terms with the ecological impact of habitat destruction, ecosystem health, biotechnological development, and climate change—as well as their broader impacts on human well-being. However, the bioethical perspectives offered in this section give us rich ways of talking about both the nature and value of Nature. Such considerations of the environment in bioethics gain more and more traction as scientists, ethicists, and policy makers come to better understand the interconnections between living things and their environment.

The Future of Environmental Ethics (2010)

HOLMES ROLSTON III

The Millennium Ecosystem Assessment gives us a stark warning about the future of environmental ethics. "Human activity is putting such strain on the natural functions of Earth that the ability of the planet's ecosystems to sustain future generations can no longer be taken for granted" (Millenium Ecosystem Assessment Board 2005b, 5). This environmental concern about the future is much like that of a former vice president of the United States who won the Nobel Peace Prize for his concern about "an inconvenient truth" (Gore 2006; Guggenheim 2006), or another once almost-president and his wife who wrote a book on today's environmentalism (Kerry and Kerry 2007). President Obama, too, has called himself an environmentalist (White House 2009). Paul Hawken, an Australian, says environmentalism, though perhaps not a single movement, as a collection of the many different organizations that are green or are concerned with a sustainable future, is larger than any other single movement in the world (Hawken 2007). The 1992 Rio Conference on the environment (UNCED), which I attended, included the largest number of heads of state ever to gather in one place. In late 2009 even the Pope made a statement on environmental concerns, dealing with various topics such as environmental justice (Benedict XVI 2009).

I think that we are at what we might call one of the ruptures of history. When Columbus arrived in 1492, this was a rupture in history for Native

Americans. Their future was never going to be like their past. Or when Copernican astronomy was demonstrated by Galileo, worldviews were going to change. Or when we heard Darwin with his evolution of species, again worldviews were going to be radically transformed. Now environmental concerns are of that order and kind.

Culture and Nature: Managed Planet? End of Nature?

What might this mean? That's what we're going to explore here. Some say we're at a rupture of history because in the future the Earth is going to be a *managed planet* and that nature is over. Putting it another way, since the origin of human beings, human technology has been contained within the biosphere. But, increasingly from here on, the technosphere will dominate the future of the planet. So far as there remains biosphere it will be managed, and it will be contained within the technosphere.

Let's go back to the thought of a visionary philosopher: Henri Bergson.

> In thousands of years, when, seen from the distance, only the broad outlines of the present age will still be visible, our wars and our revolutions will count for little, even supposing they are remembered at all; but the steam-engine, and the procession of inventions of every kind that accompanied it, will perhaps be spoken of as we speak of the bronze or of the chipped stone of prehistoric time: it will serve to define an age. (Bergson 1911, 146)

Looking back, thousands of years in the future the Second World War and the First World War are not going to seem that big on the scene. But the steam engine, the locomotive, the jet plane—power that we get from steam, petroleum, and electricity; essentially we might want to say "motors and gears"—the vast powers that we have in the modern world are going to shift an age as dramatically as the coming of fire, or as dramatically as shifting from the stone age to the bronze age. The technosphere will serve to define an age.

Some say that the future is a postnatural world. We will live in a world of virtual nature. Yes, we have nature in the parks, in the gardens, in the zoos, but it's a highly managed nature that's in some sense postnatural. Bill McKibben, who despite not being a philosopher has enormous philosophical insight, laments that we are at just such an end of nature. The character of the world is changing. McKibben claims, "There's no such thing as nature anymore" (McKibben 1989, 69).

Michael Soulé, a great conservationist, laments: "In 2100, entire biotas will have been assembled from (1) remnant and reintroduced natives, (2) partly or completely engineered species, and (3) introduced (exotic) species. The term natural will disappear from our working vocabulary" (1989, 301). An environmental philosopher, J. Baird Callicott, says "Nature as other is over" (1992, 16).

Another ecologist, Daniel Botkin, defines the problem a little differently: "Nature in the twenty-first century will be a nature that we make . . . We have the power to mold nature into what we want it to be" (1990, 192–93, 200–201). We face a new kind of nature. Botkin wants to phrase it that way so that he can include humans of one of nature's components. He wants a postmodern concept of nature with humans staying in the natural picture. Yes, he says that the ecologist must, and the philosopher must, pay some attention to the ecological impact of our technologies. Well, is the rupture that we're facing an end of nature? We might well want to ask whether most of nature is already gone, as Bill McKibben believes.

A study by a Scandinavian team on human intervention in natural ecosystems from 1992 reported that nearly half of all terrestrial nature is little disturbed (Hannah et al. 1994). But this half is mostly Saharan desert or Arctic north, not habitable nature. If you look at habitable nature, the authors say, twenty-seven percent is yet little disturbed. About a third of habitable nature is partly disturbed. So-called dominated nature makes up again some third of the landscape. You can look at those figures and say that they paint a pretty gloomy picture because dominated or partially disturbed nature makes up three quarters of the terrestrial planet. But you could also put a different spin on those numbers. You could say, "Well, the little or partially disturbed parts of nature are two thirds of the planet."

In another study Peter Vitousek and his colleagues say that humans now capture forty percent of the net productivity of the terrestrial landscape (Vitousek et al. 1986). Again you could say, "Well that's a rather dismal figure," but it does mean that sixty percent of the planet's photosynthetic productivity is still in the wild.

As we recall from our trigonometry classes long ago, an ellipse, unlike a circle, has a double focus. We can think of one focus of an ellipse as culture: culture is certainly common on the planet. The other focus represents nature, which is still present on the planet, as indicated by the Scandinavian study. The elliptical space indicates that much of the planet is a hybrid between these two foci of nature and culture. This area is symbiotic, where there is synthesis of the wild and the domestic.

The phrase *virtual nature* draws on a couple of sources. Ellen Wohl says that the Colorado River, which of course is a major river in the West, is a "virtual river" (Wohl 2001). We might imagine or hope that the Colorado River is wild and free, but she says no: it's so heavily managed that in fact every drop of water in the river is flowing where and when it flows because the water engineers have punched a button so that this amount goes through this dam here or is held back, and so forth.

Maybe the Colorado River is a virtual river, at least the lower parts of it. The river flow seldom reaches the sea. But the upper parts of the river, not far from where I live, have no dams and might still be thought to be wild. Do we want to think of Yellowstone Park as "Virtual Yellowstone"? In the East the Adirondacks are supposed to be wild and free—so the park mission statement claims—but perhaps they are going to be so managed that they'll become virtual Adirondacks. The Colorado River begins in Rocky Mountain National Park. Do we want to think of a national park, this park near my home, as being virtual? Are your local waters so managed and controlled that you can't think of them as genuinely wild? In managing nature, do we want to end nature?

Global Warming: "Too Hot to Handle?"

Think a little about what such management is doing. For instance, we might have some misgivings because it looks as though global warming is a result of our efforts to enlarge the technosphere. The result might be, well, too hot to handle. Upsetting the climate changes everything. Maybe that ends nature, by shifting what's in the air, the waters, and the soils. Climate change may shift the composition of forests, plants, and animals that can grow here and there. It may shift ocean currents and shorelines such that we may be able to grow wheat where we could not grow it before, and we may be unable to grow wheat where we can grow it now.

We just had a Copenhagen conference where the Chinese, the Americans, the Indians, and others were negotiating their national policies facing global warming. Climatologist John Houghton startled the British public by saying global warming threatened British security more than terrorists (Houghton 1994). Why? Houghton pointed out how many British people live near and depend on a shoreline. And if shorelines were to change much, he thought this would greatly threaten British security. Steve Gardiner, a philosopher, says, "Global warming is a perfect moral storm" (Gardiner 2011). He worries that this is a tragedy of an unprecedented convergence of complexities created because we are mixing nature and technology. We are mixing uncertainties at global and local levels.

You have choices to make scientifically about what the various climate models predict, which ones are good ones and which ones may need refining. You've got choices to make politically about whether the Copenhagen agreement amounts to much and what you can expect to accomplish with international agreement generally. We're worried too about justice or the spillover from the rich to the poor. The Chinese are going to think one way about this matter, the Indians another, the South Americans yet another. You have to worry about future generations. You've got to worry about the distribution of benefits and harms from global warming, whether we owe it to third world nations because we're causing problems, or whether it is charitable or benevolent of us to help out the island nations who fear for their future and wish to make plans in case sea levels rise.

Maybe the predicted effects of climate change will happen. Indeed, some are already happening today. But to a certain extent there is a lag time between actions we take to mitigate those effects now and what the outcome of those actions will be, ten, twenty, or fifty years from now. Development seems to be a good thing, and the possession and exertion of power seems to be a good thing; however, these things accumulate into something that is bad globally. In that context of doing "good" things, it's easy to deny that those "good" things might have a worse outcome than we thought. It's easy to say, "Well, tomorrow the engineers will figure out how to handle these problems." It's easy to say, "I'm not really that much at fault. I don't think I carry that much responsibility about this," or even to be hypocritical about it. It's easy to be what moralists call a "free rider." You can even cheat or veil corruption as is often the case in this country and in other nations. We have learned that there is much corruption in high places, given the conflict between individual and national self-interest. Furthermore, what is good for the Americans is often at odds with what is good for global interests.

So we face what Garrett Hardin called the "tragedy of the commons" (Hardin 1968). I wasn't altogether happy with that term when he first introduced it because he implied that all environmental problems are of that kind. But I do think that idea and that phrase are useful in dealing with global warming. A tragedy of the commons occurs when everybody does what seems good from his or her local perspective. These actions, though locally good, accumulate into a tragedy of the commons.

We have very fragmented moral agency with different people making different contributions to CO_2 emissions and global warming. Some people have virtually no power at all to do anything about global warming. Many people can act, but think it's rather senseless to act. For instance, you might spend a little extra money and buy a hybrid car and save some emissions, but your roommate will just go out and buy a bigger, higher-powered SUV and what have you done? You have just funded his increased emissions. The Danes and the Swiss people say, "We can do a lot to stop global warming in our own nation, but what difference does it make if India and China swamp us out in everything they do?"

We're in a situation in which government and business, two big influences in our lives, may force us to do things that are environmentally destructive. I don't have much choice but to drive an automobile to buy my groceries. My wife buys most of the groceries and she likes to look in more than one store. But if she drives more than six miles, the extra consumption required by her automobile virtually offsets any kind of green shopping that she could do in any grocery market within driving range.

So is there any hope? Am I simply giving a gloomy message? I worry about that. But there are, for example, over-fishing agreements that seem to have considerable international success. We've had some success—not as much as we'd like but still some—about whether we continue whaling or not. We have a Law of the Sea, and we have a Convention on Trade in Endangered Species (CITES). These have been reasonably successful. Many say that the Montreal Protocol, which deals with hydrofluorocarbons, is the most successful international agreement in human history (EPA 2007). So, maybe, we've got some hope. Global warming has proved especially difficult to get into as tight a focus as we could endangered species and hydro-fluorocarbons, and we're not sure what future lies ahead of us. Partly this is because we face what looks to be nonlinear and cascading shifts. Also it looks as though increasingly we've got the power to change the natural world. But we can't foresee the results. Even when we are able to foresee negative results, as in the case of climate change, we might be powerless to stop them.

Human Uniqueness Versus "Pleistocene Appetites"

Whether we have hope or not to a certain extent depends on what we think about human nature and what we think about human capacities to face and stop unprecedented crises. I want to juxtapose human nature against what I'm going to call Pleistocene appetites, or human uniqueness against ancient appetites. I've said already that we live in an age where we've got engines and gears, we've got enormous technological powers; and yet others say, "Yes, you do, but you've still got muscle and blood appetites." So we have ancient appetites driving vast new powers.

At this point, some will say if you're thinking about appetites that we had in our ancient past, presumably we ought to consider some love of nature since we've evolved within natural systems as hunters and gatherers. E. O. Wilson calls this "biophilia," an evolved ingrained love for our natural environment (Wilson 1984). I think this love exists at least to some extent, but it's essentially weak compared to our desires for technology and for building. Neil Evernden says humans are the natural aliens (Evernden 1993). The one thing humans do and other species don't is that they rebuild their environments, they make artifacts. Humans have an appetite, again, for wanting more, for wanting to transform their world.

Consider this idea of Pleistocene appetites. You like sweets, don't you? You like fats, you like salt. Why do you like these things? Biologists tell us we like these things because that's what we had to have ten thousand years ago to make it through the winter. You had to get all the sweets you could get, eat all the fats you could—salt was hard to get. That's also why you like sex. You needed all the sex you could get in Pleistocene times to bring the next generation into existence, because most of your children would die. So now the claim is that humans are driven by these muscle and blood appetites at a time when you can easily get enormous amount of sweets, fat, salts, at a time at which there is more sexual freedom than there was before, and we don't have any sort of natural controls on these. We're built, they say, for overconsumption. So we get fat. We eat too much. We're built for sexual desires and we overpopulate as a natural result. We're built, they say, for pushing the limits for wanting more, without controls on these primal desires. This can get us in trouble.

We have a deep-seated belief that life will get better. That's what we've been taught for centuries. That's what the pioneers thought when they settled this new world. You ought to hope for an abundant life and work toward attaining it: that's the good life. Economists think that's what it means to be rational: to maximize your powers to exploit natural resources. They say, "Isn't that a good thing? Give people more and more of the goods and services that they want." Well you may say, "Those economists want constant development, and they think that people will always want their refrigerators and they'll want technology when they can get it. They'll want more money in the bank. They'll want more floor space in their homes."

But now if you turn around and say, "Well, ok, what about the philosophers?" They may say, "We're not always sure we agree with economists, but we do think there ought to be a right to self-development. There ought to be a right to self-realization. You have a right to abundance, and everybody has a right to abundance." So if it turns out that if you have these kind of appetites pushing you, no matter whether you're looking out after yourself or looking out after your neighbor, what are you going to get? You're going to get escalating consumption, whether you're seeking your own good or if you're a missionary in a foreign land trying to teach others how to raise more corn or how to increase development.

Now analysts are asking whether humans are well-enough equipped to deal with the kind of global scale problems that we now face. Many of our institutions—family, tribe, village, nation, law, medicine, and, I think, even school and church—were designed to build local community, to make it better and more abundant. Are these institutions capable of facing the global scale issues we face now, or those that will be faced by people in the remote future? Global and long-range future concerns don't seem to grab us much in terms of what we think we ought to do or must do. They don't offer much moral impetus. So there are biologists saying that these genes, which once got us through the Pleistocene winters, are increasingly going to prove maladaptive.

What we may hope for here depends on how much we think we can rise to a wider vision of our self-interest. Various ethicists say that, locally, one can expect some charity, some benevolence. But in international relations and on the world scene, the best we can do is move toward the collective development of enlightened self-interest. Can that happen? Will that happen?

In my lifetime one of my surprises has been the European Union. When I was growing up, the Germans were always fighting the French, the French the Spanish, and the Spanish the English. The history of Europe was the history of war. But now there has appeared the European Union, which I think has collectively enlightened the national interests of much of modern Europe. I already mentioned the Montreal Protocol that has potential to offer a similar effect. We can take action in terms of global interest: we've already got some hundred and fifty international agreements that deal with environmental concerns.

I do think we are still driven by those basic gut level instincts that we've inherited from our past—what I've called our Pleistocene appetites. But I think humans are an unusual species. I flew here yesterday in a jet plane. I checked my email last night on the Internet. We have decoded our own genome and we have designated world biosphere reserves. What am I saying? I'm saying we do many things at levels that have nothing to do with our Pleistocene-inherited, native abilities. We have more vision of humans as a unique species that can transcend their own local appetites. I don't want the idea that we have innate, ancient, genetic dispositions to become an alibi for continuing our excesses.

Sustainable Development Versus Sustainable Biosphere

Think about concepts like sustainable development and sustainable biosphere. We've got a host of decision makers now saying that sustainable development is a big thing. The United Nations Conference on Environment and Development entwined its twin concerns into "sustainable development." No one wants unsustainable development, so sustainable development is likely to remain the favored model. One hundred and fifty nations have endorsed it. One hundred and thirty of the largest corporations have endorsed it. This is a good wide-angle lens that we can be oriented around or that can give us a sense of direction. We can form coalitions around the thresholds these concepts set. It allows China to do it one way, India to do it another way.

But now I want to ask what sustainable development means. Does it mean we're going to develop further, prioritizing the economy and doing what we like to the environment, just so long as it doesn't undermine the economy? Could it mean that in some sense we're going to prioritize the environment? Yes, we must have an economy, but we have to work out that economy within a quality environment.

According to the UN, the concept of sustainable development takes first priority. "Human beings are at the center of concerns. . . ." So the Rio Declaration begins, formulated at the United Nations Conference on Environment and Development (UNCED). But if you talk to ecologists they may say, "No, the best phrase is sustainable biosphere." The Ecological Society of America has said, "Achieving a sustainable biosphere is the single most important task facing humankind today" (Risser, Lubchenco, and Levin 1991; Lubchenco et al. 1991). Sustainability is a big buzzword. Environmental philosophers like Bryan Norton preach sustainability (Norton 2005). I'm asking what kind of sustainability?

There's a big problem with sustainability generally if we continue these escalating appetites. I've said that we've switched into a new epoch; some call it the *Anthropocene*, the epoch of engines and gears. If you looked at a series of graphs of GDP, you'd see a remarkable almost right angle turn right around 1950.[2] The damming of rivers, again around 1950, also takes almost a right angle turn. We used that water; we irrigated agricultural lands with it. In turn, that required the application of fertilizer. If we consider a graph of motors and gears, of transport vehicles, 1950 again marks a dramatic shift. If you look at energy use from inanimate sources, 1950 again shows a dramatic upward turn.

So does a graph of population increase. These kinds of powers, these kinds of numbers, are strong evidence of this appetite for more and more and more. We could face quite a challenge in the future.

We want more and more and more. That's a problem even if we aim for equality for all. But increasingly in the future we face the problem of the rich getting richer and the poor getting poorer. If you go back to the 1820s, essentially in the early industrial revolution, the gap between rich nations and poor nations was about 3 to 1. That gap has dramatically shifted. If you come forward to the magic year of 1950, the gap between the rich nations and the poor countries was about 35 to 1. If you come to 1974 the gap is about 44 to 1. In 1972, when the UN met at Rio it was 70 to 1. Firm recent figures are unavailable. Many think now that the gap between the rich and poor has escalated to 100 to 1. Bill Gates and a few others own more wealth than all forty or fifty of the least developed countries. The richest two percent own more than half of the global household wealth.

We need some sense of environmental justice. We have to "speak the truth to power," as the saying goes. So let's look at somebody who did. Joseph Stiglitz, Nobel Prize winner and a marvelous economist, worked for the World Bank as a result of winning the Nobel Prize and has a sensitive conscience as well. "While I was at the World Bank, I saw firsthand the devastating effect that globalization can have on developing countries, and especially the poor within those countries. . . . Especially at the International Monetary Fund . . . decisions were made on the basis of what seemed a curious blend of ideology and bad economics, dogmas that sometimes seemed to be thinly veiling special interests" (Stiglitz 2002, ix, xiii). He continues: "I was chief economist at the World Bank from 1996 until last November, during the gravest global economic crisis in a half-century. I saw how the IMF, in tandem with the US Treasury Department, responded. And I was appalled" (Stiglitz 2000, 56). For saying these kinds of things he was essentially forced out of his position.

We've learned in philosophy that power tends to corrupt. If you remember Lord Acton: "Power tends to corrupt and absolute power corrupts absolutely" (1907). We must worry about increasing power and its power to corrupt. We have to worry about those Pleistocene appetites, but that's not simply among the poor or among us who think of ourselves as ordinary people. The sort of drive for endlessly getting more can drive the rich as well as the poor. But the drive is going to harm the poor far more than it does the rich. If you worry about deriving such a claim from an economist, here's what ecologists say in the Millennium Ecosystem Assessment: "We lack a robust theoretical basis for linking ecological diversity to ecosystem dynamics and, in turn, to ecosystem services underlying human well-being. . . . Relations between ecosystem services and human well-being are poorly understood. One gap relates to the consequences of changes in ecosystem services for poverty reduction. The poor are most dependent on ecosystem services and vulnerable to their degradation" (Carpenter et al. 2006).

What's the future? The future demands that we realize that people and Earth have intertwined destinies. That's been true in the past and that's going to be true throughout the new millennium. It will be quite a challenge to work out what that entails.

Biodiversity: "Good For Me" Versus "Good of Its Kind"

One of the things we have to work out is whether, when we support biodiversity, we do so because it is good for us or good for other creatures. E. O. Wilson calls the biospheric membrane the miracle of life on Earth (Wilson 2002). But according to the Millennium Ecosystem Assessment, we've accelerated extinction rates up to a thousand times over those typical in Earth's history (Millennium Ecosystem Assessment 2005a).

Some may think it easy to say, "Well, extinctions have been going on all the time. There are just more extinctions now than ever." Yet there is a radical difference. With natural extinction there is space left for re-speciation, but with artificial extinction at a thousand times greater rate—our kind of extinction—life on Earth will just shut down. They are not many new species generated on Walmart parking lots, for example. Maybe we are bringing an end to nature after all.

We need to save things that have medical or industrial or agricultural use, but do we save what's good of its kind? According to the delegates to the Rio conference, as we earlier heard, humans are at the center of concern. The statement that is now called the Rio Declaration was originally supposed to be called an Earth Charter. But, many of the participants from developing nations didn't want to talk about an Earth Charter. We had to name it the Rio Declaration for the lack of a better name. Humans remained at the center of concern for those participating.

Certainly, we do want to be concerned about humans. We have to make important decisions about the future with our concern for human rights. From an economic perspective, you must never save the environment at the expense of the poor. They're right, of course: you can't be proud of a policy that somehow saves nature at the expense of the poor. As Joe Stiglitz reminds us, we're not proud of an ethic claiming, "The rich can win, the poor should lose." According to the World Health Organization, priority given to human health raises an ethical dilemma if "health for all conflicts with protecting the environment . . . Priority to ensuring human survival is taken as a first-order principle. Respect for nature and control of environmental degradation is a second-order principle, which must be observed unless it conflicts with the first-order principle of meeting survival needs" (World Health Organization Commission on Health and Environment 1992, 4). Human benefits, human survival, that's the first order of business—respect nature if you can merely as a second order principle. This is the economic human rights ethic.

This perspective is, of course, humane. You have to feed the people first, right? Then, if you can, save the tigers. But on the basis of my experience, such a WHO policy will always put people first. And if people always come first, there will eventually be no tigers and there will be no rhinos. There will always be people who want to put cattle in the tiger sanctuaries. There will always be people who would shoot the rhinos for their horns. So preservation must be tied to conservation: if you are going to preserve tigers and rhinos, you have to conserve those individuals and preserve their habitat. That sometimes means you're going to have to make a sanctuary for wildlife. Indeed, we have some 600 conservation areas in some 80 countries: what comes first here, at least, is wildlife conservation.

Nonetheless, we get people who say, "Well, people have to win." So we must pursue a win-win ecology, as Mike Rosenzweig calls it (2003). We have to figure out some way to keep the elephants and get the African peoples—all of them—fed: that would be a win-win ecology. On my campus we have signs that say, "Recycling, everybody wins." When you recycle, nobody loses. Some argue that the global crisis is like that. If you can get in harmony with nature, everybody will win. The best we can do is to get people saving rhinos and tigers here and there because it is in their self-interest to do so: ecotourism, for example, will in turn provide food and income, especially in a free market democracy.

Always prefer humans! And always prefer Americans first, right? Let's make that comparison. Canadians come second. The British next. You say, "That's outrageous!" Well, it may be outrageous but we have it on the authority of George W. Bush, "First things first are the people who live in America" (Brussels 2001). Outrageous? People come first and wildlife comes second? Elephants, tigers, will survive only if they're worth more to us alive than dead: if we can make some money off of them, we'll keep some tigers.

There is something naïve about one species saying it always comes first and other things can stay around on the planet only if they're good for us. Putting it politely perhaps, we are primary and they are secondary. But if we continue our extinction rate in the future we're not going to be proud of destroying this miracle we've been given.

We can't win unless we get our values right. My own ancestors were slave owners and they lost a war. But they won in the end because they changed their goals and came to see equality across race and gender. Maybe some of my male colleagues lost jobs because employers gave priority to women, but men win when women enter the workforce and become prominent citizens.

But you may think that, in saying we're winning, I'm moving the goal posts: I'm saying the goal was once winning the Civil War, but now the goal is racial and gender equality. Indeed, if you're playing football you can't move the goal posts to win the game. But if you're doing ethics, maybe you do have to move the goal posts and get them in the right place. Perhaps the next step is to say that every form of life is unique, as the World Charter for Nature says (United Nations 1982). To win we're going to have to give more respect to the miracle of life on the planet: winning requires *Earth ethics*.

Earth Ethics

I was taught to be a good citizen when I went to school. I hope you, too, are being taught to be a good citizen. But today, beyond citizenship, we need to learn to be good residents of our natural home. Isn't ecology, after all, the logic of a home? Yes, we need to be good citizens of our nations. But we have to realize that often being a good resident means something different. We must aspire to be good residents on Earth, dwelling on the home planet with resources we all need. To date, we protect those resources with our armies and with our politics. We keep nationalizing natural resources rather than thinking of them as within a global commons.

Consider two versions of the word "earth," Earth uppercase and earth lowercase. In its lowercase form, we mean dirt. Dirt is obviously a natural resource but dirt has no value except insofar as we can get bread out of the dirt. But consider the uppercase form, Earth. Earth in that sense is thought of as a marvelous planet. Maybe in fact, unlike valueless dirt, Earth is the most valuable entity of all because it generates and sustains us. It sustains life. Think about the value of life as a creative process on Earth and not simply earth as a resource that we can exploit. I own a little real estate, about a quarter of an acre and the house on it; so, yes, that earth—that dirt—belongs to me. But I want to think further, to think on grander scale. I belong to Earth and on Earth, as do we all.

At my university in a curriculum study a few years back, we asked, "What do students need to know today that we faculty didn't get taught when we went to school?" Well they need to know the Pythagorean theorem, we still teach them that. Maybe they need to know the philosophy of Thomas Aquinas, arguments for the existence of God: yes, we need to teach that to the next generation. But what's different? For one thing, we said, the next generation has got to be computer literate. But is that all that's different? No. More importantly the next generation has to be environmentally literate. I was taught to be a good citizen but no one taught me about being environmentally literate. Today, I would hope that you couldn't get out of a university education without being computer literate. But can you get out without being environmentally literate? You'll have to answer that—I can't.

Boutros Boutros-Ghali, speaking as the UN Secretary-General, closed the Earth Summit by saying, "The Spirit of Rio must create a new mode of civic conduct. It is not enough for man to love his neighbor; he must also learn to love his world" (Boutros-Ghali 1992a, 1). "We must now conclude an ethical and political contract with nature, with this Earth to which we owe our very existence and which gives us life" (Boutros-Ghali 1992b, 66–69). Well he's a politician, you may say, and politicians have to make good speeches to stay in business. But maybe he is quite right. Civic conduct might not be enough. While we're loving our neighbors we have to also learn to love our world. If you want to consider ethics in terms of a social contract, we must develop a contract with nature, a contract with this Earth that gives us our life.

If you want to switch from a politician to a rocket scientist, consider the words of Edgar Mitchell, Apollo 14 astronaut: "Suddenly, from behind the rim of the moon, in long, slow-motion moments of immense majesty, there emerges a sparkling blue and white jewel, a light, delicate sky-blue sphere laced with slowly swirling veils of white, rising gradually like a small pearl in a thick sea of black mystery. It takes more than a moment to fully realize this is Earth . . . home" (quoted in Kelley 1988, at photographs 42–45). One small pearl in a thick sea of black mystery. Maybe conserving life on Earth is the way we have to think of our roles as residents of this planet.

Another astronaut, Michael Collins noted: "I remember when I looked back at my fragile home . . . a tiny outpost in the black infinity. Earth is to be treasured and nurtured, something precious that must endure" (Collins 1980). Perhaps we are looking not back but forward to the future of environmental ethics: it's the elevation of an urgent world vision. We're searching for an ethic of respect for life on Earth. That is the future of environmental ethics.

Notes

1. An earlier version of this contribution was previously published in Holmes Rolston III, "The Future of Environmental Ethics," *Philosophy and the Environment: Royal Institute of Philosophy Supplement 69* (Cambridge: Cambridge University Press, 2011), 1–28. The author is grateful to Cambridge University Press for permission to reproduce previously published sections.

2. See http://www.usgovernmentspending.com/us_gdp_history for a graph of US Gross Domestic Product (GDP). Dr. Rolston's graphs may be found in Will Steffen et al., 2004. *Global Change and the Earth System: A Planet Under Pressure*. Berlin: Springer.

References and Recommended Reading

Acton, Lord. 1907. "Letter to Bishop Mandell Creighton." In *Historical Essays and Studies*, edited by J. N. Figgis and R. V. Laurence. London: Macmillan.

Benedict XVI. 2009. "Message of His Holiness Pope Benedict VI for the celebration of the world day of peace." *Vatican*. http://www.vatican.va/holy_father/benedict_xvi/messages/peace/documents/hf_ben-xvi_mes_20091208_xliii-world-day-peace_en.html.

Bergson, Henri. 1911. *Creative Evolution*. Trans. Arthur Mitchell. New York: Henry Holt.

Botkin, David. 1990. *Discordant Harmonies*. Oxford: Oxford University Press.

Boutros-Ghali, Boutros. 1992a. "Extracts from closing UNCED statement, in an UNCED summary, Final Meeting and Round-up of Conference." June 14, UN Document ENV/DEV/RIO/29.

Boutros-Ghali, Boutros. 1992b. "Text of Closing UNCED Statements, in Report of the United Nations Conference on Environment and Development," 1992, vol. IV, p. 66–69. UN Document A/CONF.15L26.

Brussels, James Graff. 2001. "Bad Air over Kyoto." *Time.com*. http://www.time.com/time/magazine/article/0,9171,104674,00.html.

Callicott, J. Baird. 1992. "La Nature est Morte, Vive la Nature!" *The Hastings Center Report* 22 (5): 16–23. doi:10.2307/3562137.

Carpenter, Stephen R., Ruth DeFries, Thomas Dietz, Harold A. Mooney, Stephen Polasky, Walter V. Reid, and Robert J. Scholes. 2006. "Ecology. Millennium Ecosystem Assessment: Research Needs." *Science* 314 (5797): 257–58. doi:10.1126/science.1131946.

Collins, Michael. 1980. "Foreword," in *Our Universe*, edited by Roy A. Gallant. Washington, D.C.: National Geographic Society.

EPA. 2007. "Montreal Protocol: Backgrounder." *United States Environmental Protection Agency*. http://www.epa.gov/ozone/downloads/MP20_Backgrounder.pdf.

Evernden, Neil. 1993. *The Natural Alien: Humankind and Environment*. Toronto: University of Toronto Press.

Gardiner, Stephen M. 2011. *A Perfect Moral Storm: the Ethical Tragedy of Climate Change*. Oxford: Oxford University Press.

Gore, Al. 2006. *An Inconvenient Truth: The Planetary Emergency of Global Warming and What We Can Do About It*. New York: Rodale.

Guggenheim, Davis. 2006. *An Inconvenient Truth*. Dir. Davis Guggenheim. Perf. Al Gore et al. Film.

Hannah, Lee, David Lohse, Charles Hutchinson, John L. Carr, and Ali Lankerani. 1994. "A Preliminary Inventory of Human Disturbance of World Ecosystems." *Ambio* 23 (4/5): 246–50.

Hardin, Garrett. 1968. "Tragedy of the Commons." *Science* 162 (3859): 1243–48. doi:10.1126/science.162.3859.1243.

Hawken, Paul. 2007. *Blessed Unrest: How the Largest Movement in the World Came Into Being and Why No One Saw It Coming*. New York: Penguin.

Houghton, John T. 1994. *Global Warming: The Complete Briefing*. Cambridge: Cambridge University Press.

Kelley, Kevin W., ed. 1988. *The Home Planet*. Reading, MA: Addison-Wesley.

Kerry, John, and Teresa Heinz Kerry. 2007. *This Moment on Earth: Today's New Environmentalists and Their Vision for the Future*. Cambridge, MA: PublicAffairs.

Lubchenco, Jane, Annette M. Olson, Linda B. Brubaker, Stephen R. Carpenter, Marjorie M. Holland, Stephen P. Hubbell, Simon A. Levin, James A. MacMahon, Pamela A. Matson, Jerry M. Melillo, et al. 1991. "The Sustainable Biosphere Initiative: An Ecological Research Agenda: A Report from the Ecological Society of America." *Ecology* 72 (2): 371–412. doi:10.2307/2937183.

McKibben, Bill. 1989. *The End of Nature*. New York: Random House.

Millenium Ecosystem Assessment Board. 2005a. "Chapter 4 Biodiversity." *Millenium Ecosystem Assessment Board*. http://www.maweb.org/documents/document.273. aspx.pdf.

Millennium Ecosystem Assessment Board. 2005b. "Living Beyond our Means: Natural Assets and Human Well-Being, Statement from the Board." *Millenium Ecosystem Assessment Board*. http://www.maweb.org/documents/document.429.aspx.pdf.

Norton, Bryan. 2005. *Sustainability: A Philosophy of Adaptive Ecosystem Management*. Chicago: University of Chicago Press.

Risser, Paul G., Jane Lubchenco, and Samuel A. Levin. 1991. "Biological Research Priorities—A Sustainable Biosphere." *Bioscience* 41 (9): 625–27. doi:10.2307/1311700.

Rosenzweig, Michael L. 2003. *Win-Win Ecology: How the Earth's Species Can Survive in the Midst of Human Enterprise*. Oxford: Oxford University Press.

Soulé, Michael E. 1989. "Conservation Biology in the Twenty-first Century: Summary and Outlook." In *Conservation for the Twenty-first Century*, edited by David Western and Mary Pearl. Oxford: Oxford University Press.

Stiglitz, Joseph. 2000. "The Insider: What I Learned at the World Economic Crisis." *The New Republic* 222 (16/17, April 17 and 24): 56–60. http://www.tnr.com/article/politics/the-insider.

Stiglitz, Joseph. 2002. *Globalization and its Discontents*. New York: Norton.

United Nations. 1982. *World Charter for Nature*. A/Res/37/7. http://www.un.org/documents/ga/res/37/a37r007.htm.

Vitousek, Peter, Paul R. Ehrlich, Anne H. Ehrlich, and Pamela Matson. 1986. "Human Appropriation of the Products of Photosynthesis." *Bioscience* 36 (6): 368–73. doi:10.2307/1310258.

The White House. 2009. *Remarks by the President at the Morning Plenary Session of the United Nations Climate Change Conference.* Copenhagen, December 18. http://www.whitehouse.gov/the-press-office/remarks-president-morning-plenary-session-united-nations-climate-change-conference.

Wilson, Edward O. 1984. *Biophilia: The Human Bond with Other Species.* Harvard: Harvard University Press.

Wilson, Edward O. 2002. *The Future of Life.* New York: Random House.

Wohl, Ellen E. 2001. *Virtual Rivers: Lessons from the Mountain Rivers of the Colorado Front Range.* New Haven: Yale University Press.

World Health Organization Commission on Health and Environment. 1992. *Our Planet, Our Health: Report of the WHO Commission on Health and Environment.* Geneva: World Health Organization.

Biographical Sketch

Holmes Rolston III, http://lamar.colostate.edu/~rolston/, often cited as the father of environmental ethics in the United States, is the University Distinguished Professor of Philosophy at Colorado State University. He has authored eight books and over a hundred articles and has lectured internationally on issues around value and the natural environment. He has consulted for numerous conservation and policy groups as a part of his focus on the importance of social and political engagement on environmental issues.

Climate Change, Human Rights, and the Trillionth Ton of Carbon (2010)

HENRY SHUE

Will there really be a morning?
Is there such a thing as day?
Could I see it from the mountains
If I were as tall as they?

Has it feet like water-lilies?
Has it feathers like a bird?
Is it brought from famous countries
Of which I have never heard?

Oh, some scholar! Oh, some sailor!
Oh, some wise man from the skies!
Please to tell a little pilgrim
Where the place called morning lies!

—*Emily Dickinson, "Will There Really be a Morning?"*

Those plaintive lines are, of course, written in 1859 by Emily Dickinson in Amherst, Massachusetts. She wasn't thinking about climate change then, although 1859 is only a hundred and sixty years ago. In fact, no one was thinking about climate change. We didn't really have the concept yet, although we were already shoveling coal and firing up steam engines and revving up the Industrial Revolution, which, of course, has made us rich and happy—richer anyway—and inaugurated the explosion of greenhouse gases. But it seems to me her questions are maybe appropriate for the younger people of today because their future looks darker than my future looked when I was their age.

Anthropogenic Climate Change Is Happening

My generation has failed to grasp the nettle, bite the bullet, tackle the problem of climate change. We collectively—and this of course doesn't really include me but the scientists of my generation—have done only half the job. When I was young nobody knew anything about climate change and, during the course of my life, people have figured out what it is and that people are causing it. What we haven't done is set up the political and institutional framework to deal with it, and so the task will fall to you. I think this task is urgent and what I want to do here is give you some of the reasons why I think dealing aggressively and vigorously with climate change is urgent.

I started working on climate change in 1990. Of course, we knew nothing about it when I started. We used to look up the greenhouse gases in little charts. One thing you want to know of course is what is called the *atmospheric residence time* (Meehl et al. 2007, 824–25). Climate change is being caused because some gases, which we call greenhouse gases, are collecting in the atmosphere and blocking the heat from escaping the planet. So one thing we need to know about any given gas is how long it stays in the atmosphere once it's there: this is the atmospheric residence time. In 1990 we believed that the atmospheric residence time of carbon was one hundred years and I remember thinking: "Wow, one hundred years: that's a really long time!" Now we have discovered that that was actually a wild underestimation.

I would like to share with you what was said about the residence time of carbon in the last Intergovernmental Panel on Climate Change reports. As you might know, the Intergovernmental Panel on Climate Change, the IPCC, does an updated report about every five years: the last set of these reports came out in 2007. There are always three volumes. The following quote is from the Working Group I report concerning the basic science. It says:

> [An atmospheric] lifetime for CO_2 cannot be defined . . . The behavior of CO_2 is completely different from the trace gases with well-defined lifetimes. Stabilization of CO_2 emissions at current levels would result in a continuous increase of atmospheric CO_2

over the 21st century and beyond. . . . In fact, only in the case of essentially complete elimination of emissions can the atmospheric concentration of CO_2 ultimately be stabilized at a constant level. . . . More specifically, the rate of emission of CO_2 currently greatly exceeds its rate of removal, and the slow and incomplete removal implies that small to moderate reductions in its emissions would not result in the stabilization of CO_2 concentrations in the atmosphere, but rather would only reduce the rate of its growth in coming decades." (Meehl et al. 2007, 824–25)

The Implications of Climate Change

Now that is actually, I think, quite a mind-boggling statement. It really means that once carbon dioxide is emitted into the atmosphere it stays there over any period of time that might be of interest to human beings. The more CO_2 we emit, the more CO_2 stays in the atmosphere. So, at whatever level the concentration of CO_2 in the atmosphere peaks, it is going to stay at that level for a very long time—certainly generations—something on the order of a thousand years. And this means that the duration of carbon in the atmosphere is a problem unlike almost all others, with the possible exception of nuclear waste and maybe the most persistent toxic chemicals in the atmosphere. Of course, compared to toxic chemicals and nuclear waste, carbon dioxide emission is a far more pervasive problem. The fact the atmospheric lifetime is so long basically means that what goes up doesn't come down, in the case of carbon. Now, it's not literally true that it all just stays there forever: there is chemical decomposition. The molecule does break up. The carbon molecules and the oxygen molecules do go their separate ways eventually, but only over a really long time; and so for any practical matter, any policies or decisions that humans want to make, carbon dioxide is effectively just staying there. So every addition is a net addition.

Now if you're running water into a bathtub and the water's draining out at the same rate it's coming in, it means that the water doesn't overflow the tub. But if you put a plug in the drain, then pretty soon the bathtub will overflow. And you might think, "Well ok, I'll slow down the water coming in." But slowing down the rate of the water coming in doesn't keep the bathtub from overflowing. It keeps overflowing as long as any water is coming in, and that's roughly the deal with carbon dioxide. Any amount that you add is a net addition. A higher concentration is building up, and although climate change is of course very complicated, our best bet is that the higher the peak in the atmospheric concentration of greenhouse gases as a whole—and most notably carbon dioxide—the greater climate change will be. In Bill Clinton's campaign against George Bush there was this famous slogan: "It's the economy, stupid." For climate change I think the slogan should be, "It's the stocks, stupid." It's not the emissions. Everybody's saying, "Well ok, we'll reduce emissions to fifty percent or we'll reduce emissions to eighty percent," and of course compared to doing what we're doing now, which is just increasing emissions, that would be a great improvement. But if you reduce CO_2 emissions by eighty percent, the atmospheric concentration continues to grow; it just doesn't grow as fast as if you hadn't reduced emissions that much. So, I'll repeat the sentence from the IPCC report: "In fact, only in the case of essentially complete elimination of emissions, can the atmospheric concentration of CO_2 ultimately be stabilized at a constant level" (Meehl et al. 2007, 824).

Now, I think this gives us actually two different, but each very important, reasons why action on climate change is urgent. The first one, of course, is simply the longer it takes for us to deal with the problem—that is, to stop increasing the atmospheric concentration, which we can only do by totally stopping net carbon emissions into the atmosphere—the worse the problem is going to be when it does stop. The problem is exactly like either budget deficits and national debt or fertility rates and total population: if you're running a budget deficit—if you're running a large budget deficit—and you reduce it so that you're only running a smaller budget deficit, you're still making the national debt bigger. As long as you're running any deficit, the national debt is getting bigger. You have to be running no deficit at all to stop increasing the total debt. Likewise with population, the longer it takes you to get down to replacement rate of fertility, the bigger your population will be when it stabilizes. So if you take twenty-five years, the population will be smaller than if you take fifty years, if you started from the same population. It's exactly the same thing here.

So, one reason for urgency is just that the problem is getting worse. That would be, I think, a good reason why we should act even if it weren't our responsibility at all. It would just be that it needs to be done now, and you're the ones who are here to do it—including me too, as long as I last. You're going to take over. You're here and work needs to be done. It is, I think, in that sense very much like the parable of the Good Samaritan. The Good Samaritan can say, "Why me? I didn't push the guy into the ditch. He's not my father. He's not my son. I didn't promise to take care of him." The answer is: the guy's in the ditch, and you're here. You get him out, or he stays in the ditch.

However—and this is the second, totally independent reason—we actually do have enormous responsibility, I think, because we are making things worse for the people to come. We are causing problems for future generations and causing problems with which they themselves may well not be able to deal; because, when a generation is born there's going to be some atmospheric concentration of carbon that they're going to inherit. If that concentration is bigger than it might have been, then they're going to be worse off than if it had been smaller. And heaven help them if they're born and they're still using fossil fuels: they're also going to have an energy regime that is causing additional cumulative problems. If instead we can leave them with some other kind of energy regime that isn't producing climate change and leave them with a smaller atmospheric concentration, they will be far better off.

There is no moral principle much more basic than "do no harm." So there is a lot of agonizing by philosophers like me about how much help we are obligated to give other people. If you see a child drowning, must you save it? What if there are two children drowning? OK, two. Well, what if the children keep falling in? How much do you have to sacrifice your own welfare in order to help other people?

But this isn't about sacrificing your welfare to help other people; this is about not making things worse for other people. When you're making things worse for other people, you don't ask, "Well, how much is this going to cost for me to stop?" I think what we generally think is, when you're making things worse for other people and harming them, you stop because you have no right to be harming them. So those seem to be two quite important reasons why we should take urgent action.

I grew up in rural Virginia in the 1940s—that was when dinosaurs roamed the forests! Something else that roamed those forests was evangelists. And in my town, every year, some evangelist would come and pitch a tent and try to convert us all; though we actually thought we didn't need converting since we were converted already, he was going to do it again. The basic method they employed you could describe as terror: you know, we were threatened with hellfire and brimstone and on the last night of the mission there would often be a sermon on the "unforgiveable sin." Of course the unforgivable sin was the sin so bad that it was impossible even for God to forgive you. Since the explanation of what the unforgiveable sin is was as vague as it was ominous, I can remember going home—I was eight years old at the time—and lying awake wondering if I had already committed the unforgiveable sin. If I had, I hadn't even enjoyed it! I didn't even know what it was, but I was screwed already. And I hadn't gotten started yet.

Of course as time went on and I became an adolescent, other problems occupied me, like how to deal with girls and such. So I sort of gave up on the unforgiveable sin and basically decided there wasn't any such thing. But now it seems to me maybe there really is an unforgiveable sin and maybe it's just being so good-for-nothing that you leave future generations with a terrific burden that they needn't have and that you could have prevented their having. But they do have it simply because you weren't paying enough attention and didn't exert yourself enough. It's not quite like eternal damnation but it's bad enough, I think, depending on exactly how severe the effects of climate change become. Of course the other thing is, in this case, that the penalty doesn't fall on the sinner: it falls on innocent people who are not responsible.

The Trillionth Tonne

Now I want to talk about another reason for action, which I think is a lot more interesting and somewhat more complicated and a little harder for me to quite get a grip on. Some of the atmospheric physicists at Oxford, some of my colleagues, have recently done some calculations and come to the conclusion that if we adopt as a policy goal keeping the temperature from rising more than two degrees centigrade above the preindustrial levels in 1859 when Emily Dickinson wrote "Will there really be a morning?," the crucial factor will be whether the total cumulative emissions of carbon dioxide exceed a trillion tonnes. The calculation is, if you want to keep the temperature from rising more than two degrees centigrade—by more than two degrees centigrade beyond where it was in 1859—then at the point at which you're not injecting any more greenhouse gas into the atmosphere, the total cumulative emissions since 1859 need to be no more than a trillion tonnes.

Now you can argue whether the proper goal is two degrees above preindustrial levels: that's a good topic for debate. I think that our target should be even lower and quite a few people—well, some people (it's a minority)—are quite vigorously advocating that the target be 1.5 degrees. Bill McKibben has a website called 350.org, because to stay at no more than 1.5 degrees, the parts per million of the carbon in the atmosphere needs to be no more than 350.

But that would actually mean having the atmospheric concentration reduced, so most people, I think, or more people, are thinking that a realistic target will be two degrees centigrade. So I'm going to use two degrees centigrade and a trillion tonnes as my examples because I want to try and be as concrete as I can, since some of this is somewhat abstract. But the fundamental point is that the logic is the same whatever the particular numbers are. This is, as far as I can see, right.

We get to choose how much temperature rise we're willing to tolerate. That is up to human society. We can let it go two degrees, three degrees, four degrees: that's up to us. Though, by the way, three or four degrees is really a disaster. But, anyway, it's our choice, what we take as the goal of our policy. But, given your target, the amount of greenhouse gas you can emit is not a matter of choice; it's a matter of fact. Among the amounts of greenhouse gases, I want to focus on carbon dioxide in particular. Of course there's some uncertainty about exactly how much we can emit to hit our target of two degrees over pre-industrial levels. A trillion is a nice round number. Maybe it will turn out to be .97 trillion tonnes or 1.03 trillion tonnes. I'm not saying we have precise scientific knowledge, but the best calculation at the moment is that something like a trillion tonnes of carbon is the maximum amount if you would keep the temperature increase target of two degrees.

But the logic is that, whatever you choose as your goal, there is going to be some amount of greenhouse gas, which is the maximum you can emit and still meet that goal. So you can emit more if you like, but you're going to get a higher temperature. If you emit more, you may get two and a half degrees or three degrees or whatever. It's the task of the scientists to tell us how much we can emit if we want to hit a particular target. It's for us to decide what that target should be. But once you've got the target, the way the planet works is going to tell you how much you can emit.

Now this I think also has a fairly mind-boggling implication. What it means is that the total emissions of carbon are zero-sum across all foreseeable generations. There is a cumulative total, which is the most you can emit and still have warming of only a certain amount. You might think of it as a sort of a budget. There is a budget of remaining emissions, and those are the emissions that can be produced by my generation, your generation, your children, your grandchildren, and your great-grandchildren. That one total is a total available to all those generations. So, any emitting anybody does comes out of the budget, and there is that much less for anybody else. You are competing with your own grandchildren for using up the budget of the carbon emissions.

Now of course we can say, "Of course, all right, we won't settle for two degrees, we can go on to three." Yes, we can do that, but what we can't do is say: "We are committed to keeping the temperature rise from being more than two degrees, but we'll emit more." That relationship is not up to us: how much we can emit will be dictated to us by the planet. So you have a zero-sum emissions amount that is zero-sum trans-generationally. Among other things,

I think this shows what we philosophers call the problem of trans-generational distribution is absolutely unavoidable: it is built in. You can't think about this at all without thinking about trans-generational distribution. Now I want to talk about possible solutions to this problem and their resulting implications.

Possible Solutions to the Problem

Let me set the stage with a little bit of numbers (I don't know very many numbers so I like to use all of them that I know). IPCC climate scientists point out that total cumulative carbon emissions as of a year or so ago were half a trillion, so we're already halfway to our maximum emission target. It took, according to them, 250 years to accumulate that much. So that's total carbon emissions of the previous two hundred and fifty years. On business-as-usual the other half trillion will be emitted in the next forty years—before 2050. So, obviously you can see that the rate at which emissions are increasing is very great indeed, and there is lots of independent evidence of that.[2] Finally—and maybe it's just one more meaningless statistic—but the rate of increase in annual emissions per year over the last forty years has been greater than the rate of increase in the emissions in any forty-year period for the last two thousand years.[3] Obviously, emissions are going up. We've done a lot of talking about reducing emissions, but we've done precious little actual reducing. Emissions are going up, and they're going up very fast.

The way that we're going to deal with this apparently, if we're going to deal with it at all, is with cap and trade. I'm not advocating cap and trade. There are other options—for example, I think carbon taxes would actually be a better option—but taxes take courageous political leadership, and we're kind of short on courageous political leadership. But anyway, rightly or wrongly, we're probably going to be trading emissions permits. Well, the point of trading emissions permits of course is to make it more expensive to use fossil fuels: you raise the price of the fossil fuel, and you hope that meanwhile you're making alternative energy more affordable. At some point there will be a crossover where the alternative energy gets to be less expensive than the price of the fossil fuel because the alternative price comes down and the fossil fuel price goes up because you're running out of permits or the permits are priced out of reach.

This is kind of a side or editorial comment, but people are complaining now about proposals for cap and trade on the grounds that they are "just one more tax." They just add more expense. Of course they're going to add expense! That's the whole point. The reason you want cap and trade is to make fossil fuel more expensive because you're trying to get people to stop using them (fossil fuels). If you could simply tell them to stop using it that would be good, but you can't, so the idea is that you price it higher and higher until they stop using it. So that's what we're going to do, it looks like, if we're going to do anything.

Here is one more problem to consider: there are a lot of poor people in the world and at the moment almost everyone is dependent on fossil fuel. There are a lot of poor farmers who need to run irrigation pumps that burn some kind of fossil fuel—petrol, or something—and lots of farms that need to use fertilizers that are made from fossil fuels. So we're going to solve our problem by driving up the price of fossil fuel, but we have a lot of poor people in the world that already can't afford the fossil fuel they need even at the current price. If we're not careful, our solution to the problem of climate change is going to make life impossible for poor people. It seems to me we can't, in decency, solve the general problem of climate change by setting up an institution of emissions trading that led to the emissions becoming so expensive that the poorest people couldn't afford them. There's a human rights problem inherent here. I'm not trying to make this any more complicated than it has to be, but I'm assuming people will have a right to the absolute minimum needed for existence. I'm not going to say much about that claim, but here's one place where it comes in.

Now, I'm going to simplify a lot here. We're not really going to have emissions permits being given to individual people. They'll go to nations and firms and so on. But just to keep it somewhat simple, think of it as if the proposal is everybody's going to need emissions permits including the poor, and we know we only have half a trillion tonnes of carbon left to emit per our budgeted emissions. So the question is: "What do we do about the poor?" Well, my suggestion is that the poorest[4], I think, would have to be guaranteed emission permits whether they can afford them or not. To put it in the simplest terms, some of the permits would be free, and they would go to the poorest. Now since what we are talking about actually is something like assigning the initial distribution of permits by assigning them to various nations, what this would amount to is a nation with a lot of very poor people would get more permits than a nation with fewer poor people. That would be the sort of practical implication.

OK, here's where we are now: we've committed ourselves, I'm assuming, to not letting the temperature rise above another two degrees, which means we can't emit more than a half a trillion tonnes of carbon dioxide which, as things are going now, will happen within the next forty years. We've got a fixed, zero-sum number of emissions, and at least the current poor need to be guaranteed permits at no charge so they're going to get them free. The poor in the next generation are also going to need to emit but by then maybe the permits are going to be even more expensive, and they're still going to have to get them free. And we don't know how long this is going to go on. Maybe, if we get the cap and trade going, the remaining half trillion tonnes of carbon will stretch out beyond forty years. Maybe it will take us longer than forty years because as the cap and trade system gets going, we will reduce our emissions, let us hope. We don't know for how long there will be poor people who will be still dependent on fossil fuel and who need free permits, but we mustn't allow for a total emission of more than half a trillion, or we'll overshoot the target of only two degrees temperature increase from preindustrial levels.

So my suggestion is that we have to just simply guarantee all generations of the poorest people free emissions permits because we can't create an institution that is starving people while solving the emission problem. But we also don't want to go over our limit. If it takes a very long time to fully develop alternative energy, even just the poor people of generation after generation would use up the whole half trillion tonnes of emissions permits, even if nobody else were emitting anything. And of course everybody else will still be emitting: the people with money will be buying and trading emissions permits. So in actuality the total emissions are going to be the free ones provided to the poor plus what other people can afford, and this is all coming out of the half trillion-ton budget of remaining emissions that are possible if we're to stay below two degrees. So what this does, in fact, is pit the living poor of any given generation against the future poor of succeeding generations because the time is going to come where we're running out of the emissions permits but there are still a lot of poor people who are dependent on fossil fuels and need them. We can't—and perhaps shouldn't—bring ourselves not to give emission credits to them, but we're running out.

One solution to this problem would be to say, "Oh, OK, then we'll give up on this ceiling. We'll go past the trillionth tonne." But that solution is bad for the future poor because now we're not capping the temperature. We're not capping it because we're worried about the current poor. Or the other choice, of course, is that we can really crack down on the emissions and say "Well, we're getting close to the limit, and we're going to go past the trillion tonnes soon, so no one gets free permits." But then some people won't have any fertilizer, or petrol for their irrigation pumps or whatever. In that case we're sacrificing the current poor for the future poor.

The solution? Don't get into that position. Don't let this string out that long. The solution is that *we've got to develop alternative energy soon* so that the poor have affordable alternative energy soon enough that we don't get into this position where we don't have permits for the poor who are still dependent on fossil fuel while keeping our emission ceiling, which we need to do to limit the climate change, which is necessary for the future poor and, of course, for future everybody. I've been concentrating on the poor here but the solution applies to everybody: the rich are going to have climate change to deal with too, though it's an easier problem to deal with if you're rich (although, we could also talk about how much easier it actually is).

The good news, I think, is if we really push hard to develop affordable alternative energy and push hard to keep carbon emissions as low as we possibly can, maybe we will get affordable alternative energy soon enough that for now we can let the poor who are actually alive have the emissions that they actually need without ultimately overshooting our ceiling. But the longer it takes to get affordable alternative energy, the longer it is before the price of alternative energy drops below the price of fossil energy and the less likely we are to avoid that dilemma. The positive way to view it is that we have a wonderful opportunity here to avoid creating a situation in which we pit the poor of whatever generation is alive against everybody in the future, including the poor of the future. We can avoid that dilemma if we get really serious now about developing alternative energy and cutting back emissions and greenhouse gases—especially carbon from fossil fuel—as fast as we possibly can. So, will there really be a morning? That's up to you.

Notes

1. A later version of this contribution was previously published in Henry Shue, "Human Rights, Climate Change, and the Trillionth Ton," *The Ethics of Global Climate Change*, ed. Denis G. Arnold. (New York: Cambridge University Press, 2011), 292-314. The author is grateful to Cambridge University Press for permission to reproduce previously published sections.

2. See Climate Analysis Indicators Tool, *World Resources Institute*, http://cait.wri.org.

3. See P. Forster, V. Ramaswamy, P. Artaxo, et al., "Changes in Atmospheric Constituents and in Radiative Forcing," in *Climate Change 2007: The Physical Science Basis*, ed. Solomon, Qin, Manning, et al., 129–234; and H. Le Treut, R. Somerville, U. Cubasch, et al., "Historical Overview of Climate Change Science," in *Climate Change 2007: The Physical Science Basis*, ed. Solomon, Qin, Manning, et al., 93–127.

4. A good further question is how do you draw the line marking out the "poorest." I don't have anything particularly helpful to say here about that except that I don't think it's that complicated, but it is complicated enough. Additionally, the leveling-off of world population is obviously also vital to dealing with the shortage of global capacity to absorb emissions, but this is far too complex to be dealt

with quickly here. It is important to bear in mind, however, that two factors that contribute to declines in fertility rates are improved child survival levels and improved education for women, both of which depend on elimination of extreme poverty.

References and Recommended Reading

McKibben, Bill. n.d. *350.org*. http://www.350.org.

Meehl, Gerald A., et al. 2007. "Global Climate Projections." In *Climate Change 2007: The Physical Science Basis. Contribution of Working Group I to the Fourth Assessment Report of the Intergovernmental Panel on Climate Change*, edited by S. Solomon, D. Qin, M. Manning, Z. Chen, M. Marquis, K. B. Averyt, M. Tignor, and H. L. Miller. Cambridge: Cambridge University Press. Available online at http://www.ipcc.ch/publications_and_data/ar4/wg1/en/faq-10-3.html.

Shue, Henry. 1996. "Environmental Change and the Varieties of Justice." In *Earthly Goods: Environmental Change and Social Justice*, edited by Fen Osler Hampson and Judith Reppy. Ithaca, NY: Cornell University Press.

Shue, Henry. 2001. "Climate." In *A Companion to Environmental Philosophy*, ed. Dale Jamieson. Malden, MA: Blackwell. doi:10.1002/9780470751664.ch32.

Shue, Henry. 2005. "Responsibility to Future Generations and the Technological Transition." In *Perspectives on Climate Change: Science, Economics, Politics, Ethics*, ed. Walter Sinnott-Armstrong and Richard B. Howarth. Amsterdam, San Diego: Elsevier. doi:10.1016/S1569-3740(05)05012-1.

Shue, Henry. 2010. "Deadly Delays, Saving Opportunities: Creating a More Dangerous World?" In *Climate Ethics: Essential Readings*, edited by Stephen M. Gardiner, Simon Caney, Dale Jamieson, and Henry Shue. Oxford: Oxford University Press.

Shue, Henry. 2011. "Human Rights, Climate Change, and the Trillionth Ton." In *The Ethics of Global Climate Change*, ed. Denis G. Arnold. Cambridge: Cambridge University Press. doi:10.1017/CBO9780511732294.015.

Biographical Sketch

Henry Shue, http://www.merton.ox.ac.uk/fellows_and_research/shue.shtml, is a senior research fellow and professor of Politics and International Relations at Merton College, University of Oxford. He has been internationally recognized for his work on international distributive justice, human rights, normative theory, and advocacy for ethics in the public sphere. After work on strategies regarding nuclear weapons in the 1980s, his writing has centrally concerned the issues of justice arising in international negotiations over climate change.

Ethics, Environment, and Nanotechnology (2009)

BARBARA KARN

Be advised, I am neither a philosopher nor an ethicist. However, I am pleased to have been invited by philosophers to share my thoughts on ethics, environment, and nanotechnology, since I believe it is necessary to talk beyond one's narrow groups; in this case, environmentalists or nanotechnologists. It is a rare opportunity to address a very interesting subject: the impact of nanotechnology on environmental concerns and the ethic that underlies it. The first half of the chapter is for the non-nano informed: a sort of Nano 101. The second part, for the nano informed, will cover my understanding of issues in nanoethics.

What Is Nanotechnology?

We first must clarify what nanotechnology is. Its definition consists of three parts (National Nanotechnology Initiative [NNI] 2012): first, nanotechnology involves materials that are between one and one hundred nanometers, like buckyballs, or *fullerenes*, which are nanomaterials composed of carbon atoms arranged like points on a soccer ball. Second, at the scale of nanotechnology, there are size-dependent properties. For instance, different-sized cadmium selenide nanoparticles (quantum dots) display differences in their optical properties, such as color. Third, nanotechnology involves atomic level manipulations. This area is of great interest to industry for their future development.

In addition to the three parts of the definition, the scope of nanotechnology has been enlarged to encompass science, engineering, and technology plus a variety of activities from imaging to measuring, modeling, and manipulating matter at the nano length scale. In the fall of 2006, ASTM International, a standards organization, gave a similar definition of nanotechnology. They stated that nanotechnology involves dimensions approximately between one and one hundred nanometers with different kinds of properties, covering a wide range of technologies. Nanotechnology can be thought of as an enabling technology. There is no one standard industrial code for nanotechnologies, but it permeates all industries.

When we discuss nanomaterials, we are not talking about the atomic scale. The size of atoms is measured in angstroms or tenths of a nanometer. Bacteria, in the micron range, are about 1000 nm or ten times the size of nanomaterials. White blood cells range between 2000 and 5000 nm. DNA is about two nanometers in diameter. There are some nano-size particles inside cells such as DNA and ribosomes. The nanoscale is bigger than atoms but smaller than microbes and cells.

Nanomaterials are not a single kind of material or single class of materials. For example, there are quantum dot kinds of materials, fullerenes, dendrimers, metal oxides, carbon nanotubes, and single atom nano particles. Carbon nanotubes have an interesting structure. They look like a roll of chicken wire made up of carbon atoms. If the chicken wires are lined up in a row, the carbon nanotube is a conductor; if you twist it a bit, the carbon nanotube becomes a semiconductor. As a result, one could design the electrical properties of these materials based upon their structure at the nano level. Carbon nano tubes can be put inside each other to get multiwall carbon nano tubes, which have different uses from those with single walls. In addition, there are nano clays and ceramics, plus composites where nanomaterials are put into polymer substrates.

Nanotechnology encompasses many forms, sometimes of the same material. For example, the many different forms of nanoscaled zinc oxide range from 300 nm spheres to wires about a nanometer in diameter. Different forms of the same material can occur at the nanoscale in shapes that are spherical, pyramidal, oblong, and so forth.

In addition to different forms, nanomaterials have differences in properties. The following properties could change at the nanoscale: different kinds of mechanical properties like tribology, stress or strain, or the friction caused by the material, wetting properties, thermal and chemical properties, biological properties, electronic and magnetic properties, optical properties, and so forth. This list gives a sampling of the different kinds of properties that are present at the nanoscale, which can change within that size range. We can have very different properties at this level from the properties of the larger bulk materials. These differences in properties can lead to various applications that can that be used by industry to make a wide variety of products.

To date, there have been three Nobel prizes awarded in nanotechnology. In 1986, Ernst Ruska, Gerd Binnig, and Heinrich Rohrer received the Nobel Prize for the electron microscope and scanning tunneling microscope, which enabled research at the nanoscale. Before that time we could not see atoms nor move them around. These researchers invented the microscope that allowed us to do that. Robert Curl, Sir Harry Kroto, and Richard Smalley were awarded the Nobel Prize in 1996 for their discovery of buckyballs or fullerenes. In 2007, Albert Fert and Peter Gruenberg won the Nobel for their discovery of gigantic magneto-resistance that takes place at the nanoscale. This property enables flash memory in cell phones and cameras and is only one example of a very practical application of nanotechnology.

But just how extensive is nano in our economy right now? Lux Research has been studying nanotechnology markets for the last 10 years. They expect that by 2015, nanotechnology will total over $2.6 trillion in global manufactured goods, about 15% of the total output in 2015. In 2009, nanotechnology was incorporated into $30 billion in manufactured goods, a number that doubled since 2004 and continues to increase. Venture capitalists, plus governments and corporations, have contributed almost $10 billion to nanotechnology research and development.

The National Nanotechnology Initiative (NNI) is comprised of 26 United States federal agencies in an interagency committee that coordinates the kind of research the government supports within nanotechnology. Most states and other countries have similar organizations to coordinate their research and development in nanotechnology. Governments alone worldwide spent $4.6 billion in 2005, and this number continues to increase. However, we have a long way to go when we look at geographical equity in nanotechnology research. Only three major areas of the world, the US, Asia, and Europe, are involved in nano research, and their spending on research and development is relatively even. The rest of the world is not yet much involved in nanotechnology.

The major spending and manufacturing has been in the semiconductor industry. That includes flash memories and larger (terabyte) hard drives, enabled by the gigantic magneto resistance effect at the nanoscale. The memory and storage technology displays put carbon nanotubes and other nanomaterials like indium tin oxide into flat screen displays. Other fast-moving sectors that incorporate nanotechnology include optical technologies, energy, bio and health, including medical applications. These are the areas of greatest industrial activity. Environmental applications, food and agriculture uses, and new materials are also growing sectors that make use of nanotechnology.

Mihail Roco, the godfather of the National Nanotechnology Initiative, developed an early timeline for nanotechnology. Essentially everything that has happened in nanotechnology in the United States and the world tracks to this man's vision. According to Roco, we are now in the second stage of nanomaterial development. The first stage involved passive nanomaterials such as those put into composites or sunscreens. Currently we have moved into the stage of more active, heterogeneous, nanomaterials that may serve a dual function. For example, in the environment a nanomaterial might be a point-and-shoot material in which one part of it senses some pollutant and another part is the catalyst that destroys that pollutant.

We are developing more of these active uses, and eventually they will be put together in systems and hierarchical structures. In fact, our structure, the skeletal system, is made of bone that starts with nanomaterials that are then put together into bigger macroscale structures.

The last step in Roco's timeline of nanotechnology involves designing products from the bottom up. This is an example of Eric Drexler's molecular manufacturing or, as it is now called, atomically precise manufacturing—a very green nanotechnology.

To get an idea of the products that are in the marketplace, the Woodrow Wilson Center's Project on Emerging Nanotechnologies (nanotechproject.org) has developed a database of self-identified nano consumer products. It is important to realize that there is no quality control on this database in that the products are self-identified as nano by the companies' websites. However, the database does give an indication of the extent of common products that contain nanomaterials. It lists over a thousand consumer products.

We do know that there are Intel computer chips with nine-nanometer circuitry or less. Mercedes-Benz integrated ceramic nanoparticles into the molecular structure of their paint's clearcoat. Airports in Japan have self-cleaning windows because they are coated with titanium dioxide, a nanoscale photocatalyst. The titanium dioxide helps clean up the air a bit by destroying some of the organic pollutants, and it adds a lotus effect on the surface of the window, allowing the water and dirt to run off. There are nanopants that shed dirt.

Some companies are using nanosilver in washing machines and refrigerators, since silver is a good antimicrobial. It kills bacteria so clothes stay cleaner, and food lasts longer. Hard drives, tennis balls, baseball bats, and many other kinds of uses incorporate nanotechnology . . . and the list goes on.

Incorporated as it is in such a diversity of products, nanotechnology has extended to a wide range of industries and industrial sectors. The auto industry, telecommunications, the chemical industry, the medical sector, and information technologies all utilize nano. Nanotechnology then converges with other technologies to form what some have called the nano-bio-info-cogno convergence. I won't talk much about this convergence here since it is well beyond my scope here to figure out what the heck cognitive science amounts to. But I'll mention that there is a group called the trans-humanists who look at this convergence and consider how we might enhance human beings, as nano-bio-info-cogno organisms (Bostrom 2005; Kurzweil 2000).

A Paradigm Shift in the Making

Interestingly, nanotechnology is not much of a technology. It is mainly a science, and nanomaterials are central to what we call now call nanotechnology. These are just semantic musings to indicate that nanotechnology is not a new technology in the normal sense of the term. I believe we are in the midst of a Kuhnian scientific paradigm shift (Kuhn 1970). We have the opportunity to make sure that when we make the paradigm shift, we bring the technology up right, and this is where ethics comes in.

To explain a Kuhnian paradigm shift, the easiest example to understand is Galileo. Common (scientific) wisdom before Galileo was that the sun rotated around the earth. When Galileo posited the earth rotating around the sun, it caused a revolution in thinking in the scientific community . . . and got him in big trouble with the Church. It was hard for many to make the shift, because their careers had been built on developing equations to explain how the sun was rotating around the earth. When the paradigm shifted, their way of thinking had shift in order for the science to work right. This is an example of how a change in thinking can lead to a scientific revolution. Likewise, there was a belief that diseases were in our bodies from birth. The paradigm shifted when we understood diseases came from the environment. Plate tectonics is another example of a paradigm shift when we realized that the formerly considered stable continents on the earth moved.

In the case of nanotechnology normal science would assume that, no matter how small a material becomes, its properties stay the same. At the nanoscale, however, material properties can change with the size of the material. This anomaly leads to a paradigm shift, and the worldview at the nanoscale becomes a scientific revolution where the properties of matter are now tunable dependent on size. I think this revolution is really exciting. I mean, I have not lived through any other paradigm shift that I can think of except maybe the development of plate tectonics when I was young. We're very fortunate to be living in a time when there is a scientific revolution with such potential to influence the way we all think about the world.

So what is it about modern nanotechnology that is different from prior technologies, particularly with respect to ethics? We have certainly encountered new technologies before. The 20th century was filled with invented marvels. Nanotechnology, however, sets itself apart from other technologies by considering variables that we didn't think about before. Imagine some Neanderthal transitioning from the Stone Age to the Copper age. Was he asking if copper is harmful, is it more dangerous than stone, could it be made it without pollution, is it sustainable, what about risk? It is highly unlikely that the average Neanderthal considered the consequences of new technologies. Today we do consider these questions. There is an evolution in thinking and the possibility to be proactive with nanotechnology and avoid harmful consequences.

To understand proactivity, consider the difference between a fire and a sprinkler system. If a house is on fire, the blaring fire trucks go by; the news goes out; there may be pictures in the paper, on television, and on the Internet. But if a company installs a new sprinkler system in its factory, who cares? It is not exciting like a fire. No one usually notices. Similarly, getting a toothache gets your attention more than making regular dental appointments. Being proactive is hard, because we are not moved to respond to things that do not directly get our attention. So why aren't we proactive? My examples above sort of answer the question themselves. The bottom line: it's not sexy to be proactive. The future is not now. The long-term is not here immediately, and as a result, prevention is not high on the priority list. We are not proactive in our personal lives, and usually we are not proactive in research and commercial endeavors.

Life Cycles and Green Nano

But in nanotechnology we have already learned some things that should require proactive response. For example, there could be harmful effects from these materials, like pulmonary effects in rats and mice. Maybe these pulmonary effects could happen in humans. There are toxic effects on some organisms in the environment. You can put nanomaterials into a test tube with Daphnia micro-organisms, small "water fleas," and they swim to the other side of the test tube. The growth of plant roots seem to be inhibited if nanomaterials are in the soil. Products may be harmful at certain points in their life cycle. Perhaps disposal will provide more risk of exposure to nanomaterials than manufacturing. We now also know that we need to consider the full life cycle of our products. For example, it's in the use stage of automobiles where we're getting most of the pollution problems. In the case of electronic equipment, there will be disposal problems. We begin to realize that we may be looking at the wrong part of a life cycle of a product in our consideration. If we have new materials and new applications, we must ask how to respond to new risks.

There may be ways to design out toxic effects of nanomaterials both at the design stage and in the use stage. This proactive kind of approach can be called green nanotechnology. Green nano has two aspects: one is production of nano-materials and products without harming the environment or human health; the other is production of nanomaterials and products that provide solutions to environmental or sustainability challenges.

Production encompasses making nanomaterials in a green manner using what we know about engineering design and green chemistry so we design out potential toxic effects. We can try to lower the energy usage or use less toxic solvents. We could use catalytic instead of stoichiometric kinds of reactions, make the materials greener by designing products for degradation in addition to greening up production. Green nanotechnology production has an empha-sis on pollution prevention and efficiency.

Green nanoproducts can have both direct and indirect applications for solutions to sustainability problems. Two examples of direct applications in the environment are remediation using nanoparticles or sensors to help monitor nanomaterials in the environment. An indirect application would be saved energy or reduced waste. In both cases, the full life cycle of the sys-tem is considered.

Technically, we practice green nanotechnology by following the rules of green chemistry or green engineering. In both, there is a focus on things that will help the environment and which will help sustain the planet. There is a very straight-forward translation between the principles of green chemistry and green nanotechnology.

The emphasis on green design may lead to a new direction in manufactur-ing emulating nature and how it manufactures new organisms. For example, the average mammalian cell has a diameter of thirty and fifty microns, that is, thirty or fifty times the size of an average bacterium. The nucleus is about ten microns in diameter and takes up a sizeable portion of the cell. DNA is two nanometers in diameter and fits inside the nucleus. The DNA codes RNA, and that code goes clicking along the ribosome (which is twenty to thirty nanometers in diameter) as it puts amino acids together to make proteins. These proteins are either catalysts that make other parts of the organism, or they are structural forms. Everything boils down to the cell—that little nano-factory—putting together molecules that make everything else, based on the information contained in the nanoscale DNA.

This cellular nanofactory is environmentally benign. It uses water as a solvent and simple ingredients like carbon, oxygen, nitrogen, hydrogen, phosphorus, sulfur, as starting materials. The processes take place at room temperature. There are small machines in the cell with complex feedback loops and redundancy—all in the DNA molecule. Even the end of life is accounted for. When that cellular factory dies, some critter is going to eat it up, or it's going to break down in the environment. In other words, we do not have to worry about disposing of a hazardous substance—and we do not break any laws of chemistry or physics making organisms. If nature can do this, maybe we can emulate nature, and build analogous little factories for our products, the idea of green nanotechnology.

Nature, working at the nanoscale, can accomplish a sustainable world. Surely humans can minimize their risks and likewise help accomplish sustainability through a scientific approach that emulates nature. We may be able to accomplish sustainability through nanotechnology. Some people get really nervous about the idea of nanomanufacturing, because they immediately think of nanobots and little mechanical things. However, there is no reason why we could not make a materials soup with all the ingredients in it, make some kind of information molecule, and have this molecule put together a refrigerator of some other gadget in the soup. It is not much different from what going on in cells.

Sustainability and Ethical Concern

Now that we have some common understanding of nanotechnology in general, we will consider the ethics involved. I warned earlier that I am not a philosopher. So, what approach should be used? First, it is essential to find out what it meant by ethics. The Oxford English Dictionary definition combines all parts of many other definitions and concludes that ethics is the study of "moral principles that govern or influence conduct or a branch of knowledge concerned with moral principles." There are principles within professions, and there is also a whole field of philosophical ethics that address what is right and wrong, good and evil, what is virtuous and non-virtuous. Thus, ethics deals with the values that govern conduct.

Recently, the term *nanoethics* has been coined, and we have acronyms like SEIN (Social and Ethical Issues in Nanotechnology) and ELSI (Ethical, Legal, and Societal Implications). The government uses these acronyms, too, adding to the myriad of government alphabet soups. ELSI was used in the biomedical community before nano to refer to the ethical legal societal implications of "x." Now, x = nano. There is NE^{3}LS, which is rather scientific looking with the cube in it. It stands for Nano, Ethical, Environmental, Economic, Legal and Social issues. Nanoethics even appears in Wikipedia. Wikipedia lists nanoethics as discourse on ethical and social issues concerned with developments in nanotechnology. How else would you define it? There are clearly ethical and social issues in nano.

Adam Keiper, an early editor of the journal *Nanoethics*, was not so sure the subject material of his journal was real:

> Nanoethics . . . takes as its subject a science still aborning; many of the ethical and social ills it seeks to address are mere specula- tions about the hypothetical ramifications of theoretical technol- ogies that may prove technically impossible. It is fair to say that no scientific field or technological innovation has ever faced such intense scrutiny so prematurely. (Keiper 2007)

His last sentence describes exactly what we are doing with nanotechnology. This proactive approach is something really new. However, the first part of his quote suggests we are merely spinning our wheels. Are there not really going to be any problems? What about these theoretical technologies that may prove technically impossible? For the most part Keiper refers to molecular manufac- turing, which I discussed earlier in speculating about manufacturing like cells. A lot of speculation revolves around the hypothetical "what if?"

Daniel Callahan[1] wrote an article on medical bioethics in 1973 when the field was forming. He thought that ethicists could contribute to a new tech- nology by using a three-step process. First, what are the questions that raise ethical issues? Should we put nanoparticles out into the environment to clean up a hazardous waste dump when it might kill the fish downstream? That is probably an ethical issue. Should we make a cancer drug that might kill the patient? How will we dispose of energy efficient light bulbs that contain toxic nanomaterials? These are questions that might need some ethical answers, be- cause there is value associated with them: Is it right, is it wrong? In Callahan's terms, we want to use the "rigor of unfettered imagination" to determine what the problems are and how best to solve them.

Second, the ethicist should development some methodological strategies, some kind of systematic way to think through these moral issues. They need to add the analysis and the kinds of things that philosophers do well, like logic and consistency, and adapt this analysis to the subject matter. Then, when the ethicist talks to the engineer, the engineer does not tune out all of the ethical and philosophical talk, which they might call "gobble-dee guck" (that's not my term; it's Callahan's).

The ethicist must have a concern for the particular discipline; that is, how do engineers and scientists think, what kinds of attitudes do they have, and what kinds of behavior do they associate that would need to be adapted into something more philosophical? To quote Callahan: "the ethicist should have a passion for the good" (Callahan 1973). Don't be ashamed to say that you think that good is really a great thing. You don't go in saying, "I'm neutral, so evil might be ok." It is best to support the good. Whatever that is, of course, is probably a philosophical question. We usually have an idea of what is good, and good is better than evil—sometimes only because it is easier. For example, it is easier to tell the truth than it is to lie, because you do not have to remember what you said.

Callahan's third step is to then form procedures for decision-making, to consider how these ethics can help scientists and engineers make the right decisions. He goes on to give some caveats to ethicists to try not to push a lot of philosophy on people who have a low tolerance for it: look out for philosophical overkill. In other words, if you're just trying to figure out if it should be cadmium or lithium in your product, you do not want to have to talk about the meaning of life and existence. Besides, since we already know that the answer is forty-two, there is not much to discuss.[2]

There are further issues beyond Callahan's three steps like, "Who should be allowed to decide?" But that question is largely political, not ethical. Does the boss in the lab decide? Do you decide? Does the ethicist make the decision? Does the chemical engineer or the physicist or the electrical engineer make the decisions? Ethical questions take priority; for, having answered those questions of responsible conduct, to paraphrase Callahan once more, good luck convincing these innocent scientists who think their methodology is value-free!

We constantly make choices as scientists and engineers. We can choose which direction to take our research. We can choose whether we're going to use a toxic material or something else. We can change a research direction when an anticipated result could be grossly misused. Even though scientists and engineers develop what they think is value-free knowledge based on scientific methods, they are still imposing some of their own thoughts, morals, and values onto this knowledge.

Furthermore, despite all of the discussion above, the current editor of *Nanoethics*, J. Wechert, has said that it is not at all clear that the issues in nanotechnology will be new in an interesting sense. In other words he is really saying that nanoethics may not be anything special. It may just be an extension of what we know about bioethics or animal ethics or other kinds of ethics. So the core ethical issues may not be new, but at least we are applying them early on to a new area.

Contemporary Issues in Nanoethics

So what are examples of ethically-charged issues that have been addressed in the current state of nanotechnology? First, there are Environmental Health and Safety (EHS) issues including worker protection, military uses, toxicity, exposure related to environment pollution, sustainability, and green energy. What are the good and bad aspects of these? Lawyers are extremely interested in centrally legal issues, looking for laws and regulations that could be analyzed, including intellectual property and ownership rights: "Who owns what? Should we patent? If we patent ten-nanometer titanium dioxide, can we also patent twenty-nanometer titanium dioxide?" There are also a host of societal concerns: "Who is going to benefit from this new technology? What about the North-South divide? What about the rich and poor, and economic displacements? Will we have a silicon rust belt if we begin self-assembly of our chips? Are we going to get rid of all those billion dollar fabrication plants, because we have a better way to do it? How do we educate the work force for nanotechnology? Who benefits from patents? What kind of infrastructure might we have to change? What new kind of transportation will we need?" And then there are governance issues: "What's the role of the public in governing any kind of science and scientific research and technology? What kinds of laws and regulations are there, and what are needed? What about nongovernmental organizations, think tanks, and academia? What kind of research funding is there?" The list just goes on and on. And there are groups like transhumanists who talk about changing human nature itself—extending our brainpower, extending our physical prowess.

There is a great deal of activity going on in nanoethics: there are a lot of issues we need to talk about with respect to the values at stake. This need is being met through a wide range of activities. There are conferences such as those addressing ELSI and converging technologies. The UN has put out information on nanotechnology (UNESCO 2007). The *Meridian Institute* has addressed the north/south issue and also worked on nanotechnology and the poor (Meridian Institute 2005). The *Dialogue for Responsible Nanotechnology* brought together over twenty countries to discuss how to raise this technology responsibly (European Commission 2008). There are also many university centers focused particularly on ethics of emerging technologies. The *Nanoethics* journal and numerous journal articles in scientific magazines likewise address nanoethics.

ETC was one of the first luddite types of organizations that posited a world overrun by "grey goo" as we make nanobots that are going to self-replicate and take over all the matter in the world because they need raw material (ETC Group 2003). ETC called for a moratorium on nanotechnology until we make sure it is safe. Of course, they did not get very far with that call. A lot of people are actively pursuing nanotechnology. We are still forging ahead, and we will continue. There is no way to stop knowledge or new technologies.

Private institutions like Eric Drexler's Foresight Institute and the Nanoethics Organization also work on ethical and social implications of nanotechnology. There are plenty of websites and blogs that have to do with nano and that address nanoethics in their work. Books like *Nanoethics: The Ethical and Social Implications of Nanotechnology*, and government books on ELSI from conferences also add to the conversation. There are now declarations and principles that have been developed in response to nano. The Nanotechnologies Industry Association, a trade group in England, came out with a list of principles for nanotechnologies and nanomaterials oversight in September 2007. These included mandatory regulations, health and safety rules, calls for environmental sustainability and transparency, manufacturer liability, and inclusion of broader impacts beyond just environmental. The *Responsible Nanocode* had 46 signatories as of October 2007. But without laws and regulations behind them, not much is going to happen. While I think such exercises are helpful in raising awareness, I nevertheless think one needs to be realistic about just how far a signatory document of principles or code of ethics will get.

As a case study of a moral issue, consider the element indium used in the nanomaterial indium tin oxide. Essentially the whole periodic chart is involved with nanomaterials in the commercial market. But here we will just consider Indium as one example. Indium is used mainly in flat panel displays, LED lights, and photonics. It is found in photovoltaics, in fuel cells, in solar panels, and in a wide range of consumer electronics. It is mined from zinc ores (Department of Energy 2011). This indicates a near critical problem in its use in clean energy and with a high supply risk. We have had many wars over resources. Will we have an indium war? Probably not, because we can live without our flat screen TVs, or a substitute for indium will be found. This is a dilemma to consider: give up a flat screen TV or go to war with Canada over indium.

Indium supplies exist on almost every continent except Africa. We Americans don't have a supply of indium, and we have to get it from other people in other places—and some of these places may not be politically stable. There are numerous problems here that might have ethical considerations. For instance, who gets these materials? Who is going to break the products up or recycle them? Do we know? Have we tested this? How should these limited resources be used? Can we protect workers from toxic indium? Is a flat screen TV more important that a solar cell?

There is also a huge environmental impact in extracting indium. Suppose there's a little village somewhere and someone says, "We want your indium." This stable little agrarian village was doing fine. But now the workers are in the mines, the waterways are polluted, and families are separated. There are clear ethical issues here. Can that land be protected from extraction? Should it be?

Clearly, nanoscale science is not without ethical issues and activities around those issues. With all these real and potential issues, where does the public stand on nanotechnology? In a study released in September of 2007, over fifty percent of those sampled had not heard of nanotechnology (Kahan et al. 2007). Most of us in academia probably have at least heard of nanotechnology, but we should not assume it is general knowledge. This same study reported that only about thirty percent of those surveyed had heard of it. This indicates the huge opportunity to communicate to the public about this remarkable size scale that is enabling new technologies using different materials.

Public concern about nanotechnology has moved from "grey goo" (Drexler 1986) and molecular manufacturing—little self-replicating bots taking over the world—to concerns about privacy issues, and more recently to environmental health and safety issues. New questions are constantly on the table with the environment implications of nanotechnology front and center as the current issue.

But stay tuned. Maybe there will be some big incident regarding nanomaterials that will refocus our attention. So how should scientists and engineers behave? I suggest it is with consciousness and awareness of the values incorporated into research, with conscientiousness of what scientists do and how they communicate regarding nano, and with conscience, striving generally to do no harm.

We continue to develop these technologies and their benefits can buy us time: they can slow down the rate of our non-sustainable practices. But we really need to look at our values and behavior in order to move toward a sustainable planet. Philosophers examining the ethics of nanotechnology are a key element in the new proactive approach to science and engineering. They can also be a key element to the responsible launch of this technology by working closely with the scientists and engineers who will use nanotechnology as the basis for a new industrial revolution.

Notes

1. Callahan started the Hastings Center and was an early leader in the development of bioethics.
2. A reference to *The Hitchhiker's Guide to the Galaxy* by Douglas Adams (Random House 1979).

References and Recommended Reading

Board on Sustainable Development, Policy Division, National Research Council. 1999. *Our Common Journey: A Transition Toward Sustainability*. Washington, D.C.: National Academy Press.

Bostrom, Nick. 2005. "In Defense of Posthuman Dignity." *Bioethics* 19 (3): 202–14. doi:10.1111/j.1467-8519.2005.00437.x.

Brundtland, Gro, ed. 1987. *Our Common Future: The World Commission on Environment and Development*. Oxford: Oxford University Press.

Callahan, Daniel. 1973. "Bioethics as a Discipline." *Studies—Hastings Center* 1 (1): 66–73. doi:10.2307/3527474.

Department of Energy. 2011. "Critical Materials Strategy." *Department of Energy*. http://energy.gov/sites/prod/files/DOE_CMS2011_FINAL_Full.pdf.

Drexler, Eric K. 1986. *Engines of Creation: The Coming Era of Nanotechnology*. New York: Anchor Books.

ETC Group. 2003. *The Big Down: Atomtech: Technologies Converging at the Nano-scale*. Winnipeg, Manitoba: ETC Group.

European Commission. 2008. *Third International Dialogue on Responsible Research and Development of Nanotechnology*. ftp://ftp.cordis.europa.eu/pub/nanotechnology/docs/report_3006.pdf.

Kahan, Dan M., Paul Slovic, Donald Braman, John Gastil, and Geoffrey L. Cohen. 2007. *Nanotechnology Risk Perceptions: The Influence of Affect and Values*. Washington, D.C.: Woodrow Wilson International Center for Scholars, Center Project on Emerging Nanotechnologies. *The Project on Emerging Technologies*. http://www.nanotechproject.org/publications/archive/nanotechnology_risk_perceptions/.

Karn, Barbara, Tina Masciangioli, Wei-xian Zhang, Vicki Colvin, and Paul Alivisatos, eds. 2005. *Nanotechnology and the Environment: Applications and Implications*. Washington, D.C.: American Chemical Society. doi:10.1021/bk-2005-0890.

Kates, Robert W., William C. Clark, Robert Corell, J. Michael Hall, Carlo C. Jaeger, Ian Lowe, James J. McCarthy, et al. 2001. "Environment and Development. Sustainability Science." *Science* 292 (5517): 641–42. doi:10.1126/science.1059386.

Keiper, Adam. 2007. "Nanoethics as a Discipline?" *New Atlantis* 16 (Spring): 55–67.

Kuhn, Thomas S. 1970. *The Structure of Scientific Revolutions in International Encyclopedia of Unified Science.* Chicago: University of Chicago Press.

Kurzweil, Ray. 2000. *The Age of Spiritual Machines: When Computers Exceed Human Intelligence.* NY: Penguin Books.

Meridian Institute. 2005. *Nanotechnology and the Poor: Closing the Gaps Within and Between Sectors of Society.* Washington, D.C.

UNESCO. 2007. *The Ethics and Politics of Nanotechnology.* http://unesdoc.unesco.org/images/0014/001459/145951e.pdf.

Biographical Sketch

Barbara Karn, http://www.nanotechproject.org/about/leadership/senior_advisors/barbara_karn/, is director for the Environmental Health and Safety of Nanotechnology Program at the National Science Foundation and Senior Advisor at The Project on Emerging Nanotechnologies. She holds a PhD in biology and environmental science from Florida International University and has consulted in different capacities with several federal agencies including the Environmental Protection Agency. Karn's research and administrative work focuses on the ethical, legal, and social implications of developing nanotechnologies.

BIOTECHNOLOGIES

In the opinion of Bill Joy, a well-known computer scientist and co-author of the Java computing language, the new universe of biotechnologies—including reproductive genetics, synthetic biology, biobanking, and nanotechnologies—is leading our world into a new era in which humans may be rendered obsolete (Joy 2000). Do we take such a warning seriously or do we take his opinion simply as a new form of technophobia? Either way, Joy's concern has provoked the undertaking of novel modes of inquiry and interdisciplinary research in order to gain a more solid understanding of the impact biotechnologies will have on the well-being of the natural world, from human and nonhuman animals to the environment more generally.

You do not need to be a practicing scientist to see that emerging biotechnologies will have enormous implications for our future and that the promises of science are irremediably changing our world. Biotechnologies challenge the traditional boundaries between the human, the animal, and the machine. We are increasingly living under "blurry" conditions that as yet fail to offer any social consensus or reassuring ethical guidance. The risk is that this blurriness denies us the ability to be proactive in light of new technologies or new contexts, only allowing space to *react* to bad consequences, accidents, and negative impact.

Perhaps no specific research area can produce sufficient answers to the challenges raised by these novel biotechnologies. Lacking such answers, policy-makers cannot clearly define social policies, ethicists cannot draw helpful guidelines, and bioengineers cannot take proactive relationships to potential implications of future work. The perspectives included in this section, coming both from bioethics and from science, challenge us to think through some of these potential implications and to gain a better understanding of the current state of emerging biotechnologies. If proactive responses to developing biotechnologies *can* be developed, efforts like these are surely required. These critical perspectives on how biotechnologies may alter our lives frame a positive direction for future work in policy, ethical dialogue, and scientific development.

References

Joy, Bill. 2000. "Why the Future Doesn't Need Us," *Wired* 8.04. http://www.wired.com/wired/archive/8.04/joy_pr.html.

Nanotechnologies: Science and Society
(2007)

JAMES LEARY

As a nanoscientist, I initially wasn't sure I had anything useful to say about the broader implications of this developing technology. But, after some reflection on the subject, I have come to see that we scientists are well placed to consider the implications of our work. What I'm going to try to do in what follows is give you an overview of key issues and try to put things in perspective. Nanotechnology will have a pervasive and disruptive effect on the world around us and on human ethical (or unethical) behavior in how we apply it within our society.

The field of nanotechnology has had a somewhat unusual discourse between some of the participants involved. Through this discussion, when you see the breadth of discussion play out in various fashions in the literature you might be better able to understand where it's coming from and how to calibrate it properly. One such discussion is from an interesting book called *Nano-Hype* by David Berube (2005), who happens to be an English professor. The National Nanotechnology Initiative (NNI) requires that a certain percentage of funds be spent on discussing and researching the societal effects of nanotechnology and Berube, who is very closely associated with the University of South Carolina's nanotechnology center, serves in this capacity.

I'm going to tell you a bit about the effects of nanotechnology discussions and hopefully try to put the hype in proper perspective for you. Then, when you see things thrown at you in the news media, you'll know basically what to take with a grain of salt or what to take a little more seriously. I approach this topic from a variety of perspectives.

In my scientific life I do a lot of different things. I've been working on nanotechnology and particularly bionanotechnology for more than a decade, but I approach it from a diverse scientific perspective originally as an engineer, physicist, and biologist. While I was at MIT I also did a degree in philosophy, in particular in ethics; so, as I've said, I've lived my life by my philosophy degree and I make my living by my engineering and science degrees. In what follows, I am going to give a brief overview of several central topics around nanotechnology in the hopes of helping you think critically about them. I will separate a little bit the science fiction from the science along the way, but all of the areas I'll discuss are already impacted by nanotechnologies. I'll conclude with a topic close to my own research into which I'll go in deeper detail.

Nanotech—Hype and Reality

Nanotechnologies have been said to be the next big technological wave hitting the world. In some ways that talk is mere hype, as it is with any new technology. But I would indicate to you that there is also a lot of reality to it. It will indeed fundamentally change the way that we live, work, and think about things.

Since nanotechnology will impact on our daily lives it will also require us to be responsible for its ethical effects on humankind. We might call this area *nanoethics*. Ethics involves steering various forces in our society in hopefully a positive, meaningful way. It is the hope that we develop things constructively and with good sense, regardless of any individual's theological or political beliefs. Ethics is supposed to transcend those individualistic thoughts. As a bionanotechnologist, it is my responsibility to see that the impact of my work, and the field in general, on society is positive.

Let me give a brief introduction to some of issues among the warring factions in this field of nanotechnology. As always, science fiction leads the story—we usually get the Hollywood version before we get the scientific reality. What has actually happened in this field is quite interesting because a very famous eminent physicist, the Nobel Prize winner Richard Feynman ([1959] 1992), gave a very famous speech that was at the time quite profound. He basically said in that speech, "There's plenty of room at the bottom."[1] By that, he proposed that someday we might be able to do atom-by-atom assembly of structures. Now this was, at the time, as profound as it was really far out there. But Feynman was a very brilliant man and when you are a Noble laureate people tend to listen to you, despite how far out there you may be. He also indicated that there was no technology whatsoever in that time frame, 1959, that was able to do what he suggested. However, he planted the idea and people began to think about it. It was an amusing topic area at first, but gradually all of this "nanotechnology" started to develop.

This story is fundamentally different because throughout human history we've sculpted big objects into smaller objects. This has been the fundamental way that we have made things. In fact even in the semiconductor industry we make computer chips by photolithography. We have used light to make things smaller and smaller. We get finally down to X-rays to make the wavelengths short enough to make things even tinier so we can put more transistors on silicon chips. But we're still taking about a top-down big-to-small approach. If today, however, you visit the Birck Nanotechnology Center at Purdue or, for that matter, any nanotechnology center in the world, you'll see that many scientists are building things from the ground up—atom-by-atom. We really can do that now—we have proof-of-concept. What we're talking about here—and this is a current phrase in the literature and in the media—is molecular manufacturing. The idea is that you can physically take nanomaterials, atom-by-atom, and build everyday objects out of them.

In Feynman's wake some interesting people came along. One was an MIT student who was intrigued by Feynman's idea of atom-by-atom assembly. In 1991 this student, Eric Drexler, completed his Ph.D. dissertation and received the one and only degree in molecular manufacturing from MIT. He actually wrote what is now a very famous book in 1986 called *Engines of Creation: The Coming Era of Nanotechnology*[2] (Drexler 1986). As is frequently the case with pioneers, no one knew what to do with this. Fortunately Marvin Minsky, one of the fathers of artificial intelligence, was Drexler's mentor and managed to set what was the proper tone: "There is surely something brilliant about this, despite that we're not sure what to do with it" (see Minksy 1986).

Unlike conventional batch analysis which rearranges atoms though chemistry, Drexler basically said, "We're going to position atoms one-by-one and we're going to form structures and make anything we want." Chemists manipulate atoms now and they do it with a lot of ingenuity, but can they play with atoms as physical objects and move them around? That's hard stuff. If you've ever made things by chemistry, you really have to do a lot of things directly, indirectly, moving things around in funny ways taking advantage of electrical forces and chemical bonds. What Drexler was proposing was that you could actually physically move atoms around and put them exactly where you want to put them. People have now done this, but they have done this only on very, very simple structures. The rub here is that Drexler believed we could do this with much more complex structures. Although right now Drexler's belief is not yet fully realized, there are certainly people working seriously on the idea. While such engineering is presently science fiction, I wouldn't write it off. If it happens, and I think it eventually will, it would have a revolutionary effect because we could combine the information systems that we have now and literally send coded information out to remote places and manufacture almost on the spot. That paradigm shift would be profound. We would not need centralized factories. We would make things wherever and whenever we needed to use them. But again, this is presently science fiction: maybe it will happen, maybe it won't.

The protagonist in this debate about the possibilities of nanofabrication was Eric Drexler. The antagonist was Richard Smalley, who won the Nobel Prize for making C_{60}, or buckminsterfullerene, the soccer ball type atomic structure that you've probably seen before. Rick Smalley and Eric Drexler went into a very long protracted war of words with each other that was published in the literature. It got quite heated (see Baum 2003).

Smalley basically said that assembly by moving things around, atom by atom, faced three problems. The first problem was what he called big fingers, meaning that it's too hard to make things—fingers—small enough to manipulate atoms. Second, even if you did manage to make such tools, you'd pick up an atom and couldn't get rid of it—the problem of sticky fingers. The third part of Smalley's argument was that you couldn't do it fast enough. Eric Drexler replied. Their arguments went back and forth, as all ideas do in science. An idea comes out, maybe it's not quite there, criticisms come back, and the idea is refined. Drexler has refined his idea over the years by doing self-assembly of atoms up to a certain point and then taking the larger objects of those and then putting those together in macro-assembly. Now it's pretty clear that nanoassembly is possible.[3] In fact, living things self-assemble complex nanostructures all the time and our bodies are full of nanomachines that are responsible for many of the fascinating properties of living things.

Out There in the Marketplace

In the meantime you always know something is real when people start investing lots of money in it. This is already happening in the US. The US is currently putting out more money in nanotechnology than all of the rest of the rest of the world together, although that's changing. China and India and Europe have started pouring money in also, as evidenced by the bipartisan economic committee study that was put out by Jim Saxton (Saxton 2007). Basically, they say in the report that the nanotechnological revolution really is coming, and you can ignore it or maybe you will be able to live your life successfully without having to deal with it; but, chances are you are going to deal with it in the work place, in your education, or at some point in some aspect of your life.

The National Nanotechnology Initiative (NNI) has set out four main goals. First of all, we have to have a world-class research and development program. If we pin some of the hopes and future of America on this, then we will certainly have to conduct world-class research in nanotechnology. Second, we will have to facilitate nanotechnology transfer into real products, and economic jobs. Third, we have to develop education resources and a skilled workforce in this, and there are a lot of university courses springing up to do so in nanotechnology, nano-medicine, nanofabrication, et cetera. There are a lot of other courses in nanotechnology that are starting at the graduate level, but we have to integrate instruction much, much earlier than that. Education on nanotechnology needs to take place in elementary school, much as education in computers has occurred at this level. Lastly, we have to support responsible development of nanotechnology. I'm going to say quite a bit about what's responsible and what isn't.

Some institutions and labs already have modeled structures available as literally hands-on toys for school kids to learn about C_{60} buckminsterfullerene and its elongated form "carbon nanotubes" (the new wires of the future), as well as examples of other atomic assemblies. You'll hear more and more about things like gold nanoshells and quantum dot nanocrystals. People have even made little structures and objects. If you really want to have some fun, search for "Jim Tour at Rice University." Professor Tour's goal is to show that you can make functional nanoscale objects and make kids excited about nanotechnology. He's made a complete "nano car" complete with atomic-level wheels, axles, and a functional nanomotor that will power the car under light. He also organizes a whole series of cartoon characters to help kids to learn about nanotechnology. Tour has done a fabulous job with nanotechnology education.

The educational initiatives of the National Nanotechnology Initiative are not the only evidence of the current push toward nanotechnology. The UN and UNESCO have released a list of all the many areas of nanotechnology that are going to lead to the betterment of humankind—and it's really only a very partial list: energy (storage, production and conversion), agricultural productivity enhancement, water treatment and remediation, disease diagnosis and screening, drug delivery systems, food processing and storage, air pollution and remediation, construction, health monitoring, vector and pest detection and control (UNESCO 2006). There are fabulous things that can be done with nanotechnology in energy production and conservation. All the LED lighting that you see from your traffic stop signs to your latest home flashlights comes from nanotechnology. This technology is already pervasive. It uses less energy, it's easier to manufacture, and it has an almost unlimited lifetime. Nanotechnology can also be used to clean up water. There is much research being done to filter water and to help basically make seawater into freshwater. As you know we live in a planet two-thirds filled with water and yet we have an acute freshwater problem. So if we can come up with ways of using nanotechnology to be able to solve the water problem, it would be of vast importance. So there is a real sense of hope that this is a science and technology that can really make a difference in the world and in our daily lives. In terms of ecological impact on the planet, nanotechnology represents an important new paradigm shift. We will eventually use a small fraction of the energy to run many things in our daily life. If we think about one component of bioethics as the responsible stewardship of the planet and its resources, nanotechnology will rank very high in that category—one of the true cases of "doing more with less."

But certainly anything we create can be used for good or for bad. Some people are already spending a lot of time thinking about this. In terms of managing risks with nanotechnology as with any technology, we can do damage with it if we use it the wrong way or for the wrong things. Fire was a great thing for humans to acquire, but it also burns down buildings and harms people. We have to develop new technologies under the right set of circumstances. We have to minimize the risks and maximize the benefits. People are working very hard on this in a proactive way and the attitudes about the importance of risk analysis in this case are quite a bit different than with technologies in the past.

As many of you know, genetically modified (GM) foods have not exactly taken off in Europe. It's something that people in the US deal with on pretty much a daily basis, although certainly not everybody agrees. Europeans generally have not accepted GM foods; they don't like the way the technology was presented. They have a lot of inherent concerns about it. What I'm saying with this example is that social attitudes toward technology matter. As we all know, just because we can do something doesn't mean we should do this thing, and if we do it we have to do it in a way that manages the risk. On the other hand we need to understand the actual benefits and risks if we are to make responsible decisions about the use of technologies in our society. Unintended consequences can have serious effects.

Now, as in the case of GMO technologies, Europe has been out in front of the US in terms of thinking about some of the problems of nanotechnology and its potential risks to society. This is interesting because much of that thinking has been happening during conferences in the US. I think and hope the attitudes that are developing in the US are going to be influenced by thinking imported from Europe. But we in the US are primarily driven by economic competitiveness. If we worry about nanotechnology it's because we know that very stringent controls may lead to a loss of jobs. We're balancing loss of jobs versus impact on the environment or on human health. We're all balancing risk on some or another scale. There is no risk-free thing that you can do. In the case of nanotechnology, what I think you're going to see is the US taking more of a chance balancing these risks. You're going to see a lot more discussions of how to balance the risks of nanotechnology. And I applaud that. I think that's something we really need to do. Scientists need to bring some of these new technologies and new innovations out to the public so that we are all forced to have discussions and make intelligent decisions.

We need to have an educational process that starts very early in our children's lives. I mentioned that Tour's group has proven that you can teach the basic concepts of nanotechnology like questions of scale via some rather well done cartoons. Teaching kids about nanotechnology is very interesting because it's not just about questions of scale. When you get down into the nano world, physical laws and the things that you take for granted in everyday existence aren't the same. It's difficult to teach anyone—especially kids—that the world around you really doesn't behave the way you think it does; rather, at the macro-level, all of the differences sort of average out. But when you go down to the level of nanotechnology, things behave quite differently and sometimes strangely. When you try to teach kids, you quickly find that learn experientially: they touch things, they feel things, they smell things. But how do you teach nanotechnology when it's hard to see, touch, feel, and smell?

Since most people will not go on to graduate school and become working nanotechnologists, we have a responsibility, not only in public schools but also at undergraduate level, to teach about nanotechnology and about its effects on society. We're all going to have to make decisions about nanotechnology and its impact on our society. We don't need knee jerk reactions or fear driving these decisions; rather, we need to make intelligent decisions.

This broad education is beginning to happen, not only about the technology itself but the way of thinking. Working scientists find that they have to reformulate or retrofit their thinking and, if nothing else, to give it the right buzzwords and the right perspective. Nanotechnology is not just about being small. It's a totally different way of looking at the world. To "get" it, you have to turn yourself around and look at the world from another point of view. Working scientists in academia and in a wide range of professional fields have to be retooled. Lastly, the general public must also be retooled, because they're going to vote. We're going to have resolutions coming up that you're going to have to vote on in your home states and across the country regarding nanotechnology. The goal of nanotechnology education is to help everyone make good decisions about the future and impact of this technology. We need to make good decisions, wise decisions.

Nanotechnology is already very real. When you go shopping at the local mall, in your stores you're encountering all kinds of nanoproducts right now, more than you may imagine. Some of them will put out buzzwords that they're "nano-" this and "nano-" that. Some of them won't advertise nanotechnology even if it is present, and there's a debate about that. Right now the word "nano" is used to give products a sexy connotation. If you remember back when the computer revolution took over we used to call things "mega," and then it was "giga." Well now we're going the opposite direction, making things very small. You'll see a lot of things called "nano," some things that have absolutely nothing to do with nanotechnology. But there are in fact many things that do actually use nanotechnology in our daily lives.

Here is a list of a few things so that when you walk by a store you'll recognize the pervasiveness of 'nano': personal products (home pregnancy test by Carter & Wallace, nanocosmetics by SongSing Nanotech), household goods (antibacterial kitchenware by Nano Care Tech, Eagle One Nanowax by Eagle One), clothing (Dockers Go Khaki by Dockers), and sport/recreation products (AccuFlex Evolution Golf Shaft by Accuflex, Atomic Snow Izor Skis by Atomic). Of course, I'm not endorsing any of those products. But they are pervasive. If you go to clothing, you can buy Eddie Bauer's swim trunks, or you can buy Docker's pants, both advertised to shed water. Does that make sense? Yes it does, because most nanomaterials are very hydrophobic. They shed water. They don't allow water to sort of soak in. This is a big application of putting nano into clothing. If you put small nanoparticles into clothing, clothing doesn't wrinkle as badly. Likewise, nanoparticles can help prevent staining. We may well end up with clothing that you don't have to wash anymore! You're probably smiling now, thinking that unwashed clothing probably smell pretty awful after a while. Well guess what: it turns out they put silver nanoparticles embedded into the structures to kill bacteria. So they put silver nanoparticles in some slippers so you can buy antibacterial slippers.

Of course, there are also products masquerading as "nano" products, but real nanoproducts are only going to become more prevalent. Children might want a "nanobat" this year for baseball, even if they don't call it that. Now is that hype? Is that real? It turns out there are in fact a lot of sports products where they are embedding nanoparticles into the fiber structure, particularly in fiberglass. In bats, this actually gives much greater strength by adding very little weight so you can actually hit a ball further the more and more of that structure you make it with. Remember you can theoretically make nanoobjects that are one hundred times stronger than steel at one-sixth the weight. If you can swing something faster because it's lighter, you will actually perform better with that. You're going to be able to hit harder given the physics involved. So there is something to these children's request, even if a little bit of it is hype. Most people may not notice much of a difference, but it does make some sense from the physics point of view.

There is a nanomaterial that can fill in porous spaces in other materials and make them stronger. Other materials tend to shed water or change color. This is something neat. How would you like to have clothes that change color with your mood or if you go into a different environment? This is one they advertise and actually make: it is a physical product. There's a dress with flowers on it that open and close according to the wearer's mood, which is body temperature and other things of the person wearing it. You've seen something like this in earlier things where you put your hand on something like a ring, and the temperature of your hand tells you things about your personality or whatever. Think about having clothing that can change color. We're going to come back to that concept because guess who really wants clothes that can change color? The military. They want soldiers to blend into their background. You can imagine that if you're in an environment that's changing and you want your clothing to change right with your environment to be able effectively blend right in, the military would serious be interested in. They're actually doing it. They're building super soldier outfits that are basically not just an armor suit but actually a first line medical defense. This stuff is real. These outfits have built-in sensors that can literally monitor the soldier's body and respond to things in the field via a system of radio communications. More and more, your clothing is going to be very functional rather than merely decorative. It's going to have all kinds of functions. You don't tend to think of that right now, but that's an awful lot of material to be carrying around that doesn't do much except cover your body. It has a lot of other functions. The military is also very interested in making things like this very lightweight. We used to have knights in shining armor. Well, we're going to have soldiers in non-shining carbon nanotube armor to protect against things like shrapnel. In fact MIT has the first institute for such work: the Institute for Soldier Nanotechnologies (MIT 2012). And they're in a consortium with industry partners to actually develop and apply this sort of technology to germ resistant fabrics, enhanced medical sensors, and so forth.

You'll also see a lot of nanotechnology in cosmetics. I have a mixed feeling about cosmetics because this is the industry that in the early 1900s used to put radium in their lipstick because they thought it was really cool to have women's lips glow. There are a few downsides to cosmetic applications of nanomaterials. One of the downsides is that if you get really small nanoparticles down to the level of twenty nanometers in diameter—that's a pretty small nanoparticle you find in some of the cosmetics—they will not only go on the lip, but they will also go right through the skin, and right through the blood/brain barrier. That is scary stuff! Therefore, I advocate rigorous safety testing.

Safety and Long-term Impact

There is no regulation yet: they're still trying to figure it out. I gave a talk in 2007 to the FDA, which was then still trying to figure out how to regulate all these nanoproducts, particularly where they impinge on human health. The FDA has wisely partnered with NIST (National Institute for Science and Technology) and NCI (National Cancer Institute) to combine their respective expertise to assess the potential dangers of nanotechnology.

The size of these particles matters a lot. Twenty nanometers is a critical size where these particles can go right into your body through the skin. You think of your body as a solid thing but, the walls around your body really have a lot of holes in them when you get down to the nano world. If things are small enough, they'll go right through. When you get to larger scale particles, there seems to be less health risk. For example, some carwax formulas include nanoparticles. You can by these products at the local stores. It makes for a shinier finish on your cars. It really does work. But those particles are about two hundred nanometers in diameter so they won't go up through your skin. I wouldn't eat the stuff, but I think you're safe polishing your car with it.

Those are all great things with huge potential. But then people start to worry. What happens when you make such things? I can remember about ten years ago the amount of material actually made with nanomaterials was so small that it really didn't have any impact on the environment. Well, like most things that humans do, we quickly scale up. And, as you know from most other spheres in this world, we humans have a significant impact on the environment. Well, we're making enough nanomaterials right now that they really can have an impact on the environment.

We have to think about the cradle-to-grave impact (see Nanotechproject.org 2007) when we make something: what is going to be the environmental impact of that product? It has to be safe, responsible, and it has to be sustainable. Otherwise we're going to end up with the problems we have with nuclear waste, a result of the production of nuclear power. OK, that's great—it's cheap energy, but we've got this waste product that sits around for thirty thousand years. What are we going to do with that? We have to think about these things as we go. And if you think about it, this is what's happening. Markets right now, worldwide, predict that nanotechnology products are expected to grow from thirty billion dollars in 2005—so already not a small industry—to two point six trillion by 2014. That's astounding. That says you're going to see nanoproducts everywhere.

So there's potential for a huge environmental impact. As I said, Europeans tend to worry about quality of life more than Americans do, so they've been worried about this for a long time. If people get a particular exposure in their homes, what happens to their water, their air? The environment around them? To give you an example of that, of how we don't think about this: all of you have used nanomaterials for a long, long time because the largest use of nanomaterials in the US and the rest of the world probably has always been, until recently, paint. Paint is basically a collection of things, including nanoparticles. This was brought to bear in the context of the international Space Station: they wanted to make one of the Japanese components look good so they painted it before they brought it up. They got it up to the space station and all of a sudden the nanoparticles in the absence of gravity really leached out of the walls and were everywhere. Nanoparticles come out of the walls here, all around us. So you have to think about this: we're putting nanoparticles everywhere. Hopefully they stay where you want them to, but what are the effects of these things? You certainly wouldn't stick your head in a bucket of paint and breathe the fumes, or drink it. But you're breathing that in all the time anyway because paint particles, nanoparticles, fleck away. Small nanoparticles are too small to be pulled down to the ground by gravity. They stay in the air and get buffeted around by molecules in air.

There was a conference that was held in 2006 on recommendations for lifecycle assessment (NNI 2007). One of the problems with nanotechnology, and all new technologies, is we don't have all the data in. How do you make rules about things on which you don't have all the data? Well you can do this in various ways. You can either say, "Let's approach this thing very conservatively.

We won't let them do anything until they can prove that it is absolutely, positively safe." Well, if you know anything about science you know that scientists never prove anything. All we do is show one case where it is not safe. We cannot prove something in positive sense, because there is always another case we haven't looked at yet. The same thing is true when you're thinking about this. And you could say, "We can't wait until we have perfect data. We have to be modest about uncertainties. When we say we don't know something, just say it, live with it, deal with it." Human beings can deal with uncertainty. What you do is you build your case as you go along. Now, what I would indicate to you as you soften those economic drivers is that nanotechnology is taking off at an exponentially steep rise. It's going very fast, so if we're going to regulate we have to run as fast as we can to try just to keep it. You might say "Gee, why don't we just sit and study the problem for the next ten years?" I'm going to show you some perspectives that say why not only we're not going to do that but that the rest of the world is not going to do that either.

The idea is that you have to make what we call "provisional decisions," and scientists do this all the time. We always assume that what we know today may be wrong tomorrow. In fact we never prove anything: we just get a little better at it each day, and that's how we publish the next paper because it says something a little better than what was there before. That's all that we ever do. We have to approach issues in nanotechnology in the same kind of vein.

Now, we go one step further when we talk about nanotechnology and the environment. We can get very personal about that because most of us have to work for a living. Because of that, we're going to have nanotechnology in the workplace. We've dealt with a lot of things in the workplace in the past. We have an entire chemical industry; we have all kinds of industries that have dangerous things. We have a whole biotechnology industry and the semiconductor industry. There are a lot of hazardous things out there, but you learn ways to deal with those safely. There's a government agency called NIOSH, which is responsible for setting the safety standards in the workplace. NIOSH has been very busy and they've also been talking a lot with their European colleagues and trying to set new standards. NIOSH has set up a nanotechnology research center and is trying to examine some of these issues.

There are at least four areas to help determine if these things really pose a risk for work-related injuries and illnesses: hazard identification, exposure assessment, risk assessment, and risk management. We can make a very safe environment. You all know that, but accidents happen. What do you do when an accident happens, how do you deal with it? We have laboratory accidents and we have to have ways of dealing with those. Usually we have a happy outcome, but sometimes we don't. We need to conduct research on the application for the prevention of work related injuries and illnesses. So that's pretty important. We need to promote healthy workplaces through interventions and recommendations; we have to set some standards. Now those of you who deal with biology know that we have bio safety inspections and all kinds of provisions. We have to do things right or they will come into our labs and shut us down. We have to be very strict and there's a reason why we do that. In fact until very recently, universities were, frankly, one of the most dangerous places to work. NIOSH did not really consider their purview to be educational institutions they considered a work place. They were not really worried about academia. But now they are, and they're starting to clamp down.

We also have to think about enhanced global workplace safety. This is a big issue because countries like the US tend to sometimes say, "Well we're not going to manufacture in the US but we'll do it in Mexico" or, "We'll do it in a third world country somewhere." We've had a lot of those issues with regard to clinical trials where people have tried to do end runs around the system, do a clinical trial in a poor country and do what we consider unethical things to people. We don't want the same thing to happen in nanotechnology. People don't like regulations and rules, but, in fact, they are a way of leveling the playing field;the only reason it's cheaper to do work outside of the US is that you don't have to follow the same rules. If you have to follow the rules everywhere, then the price starts to come up and express its true cost. This is something we haven't done very well for workplace safety, but this is something I think they're trying very hard to do with nanotechnology.

Benefits of Healthcare: Where Reality and Science Fiction Merge

The area of nanotechnologies and healthcare is another area that I hope is one of those beneficial areas providing a good use of nanotechnology. And we've come a long way from in healthcare from Da Vinci's *Vitruvian Man* to non-invasive full body digital imaging (e.g., CAT, MRI, and PET scans). But we still have very far to go. First of all, there are demands for nanotechnology applications in US healthcare. You're going to see nanotechnology everywhere and a lot of it is going to be in home products. One of the products you can buy off the shelf is a home pregnancy test working with nanoparticles. Actually, nanoparticles are very sensitive so that when they bind to things they can change color. So it's like a dipstick type test, and you've seen a lot of those. Those are very nice to make out of nanoparticles on a material, and if the particles bind to a particular hormone or a protein in a solution, it will light up a different color.

The cost projection for such products is tremendous, about six and half billion dollars of nanohealth products in our system (Freedonia Group 2005). This is a very realistic and probably very modest figure compared to what it will be in a few years. It's going up extremely rapidly. As I said, anything for which you want to have a traditional rapid test—urine, blood or even saliva—nanoparticles will sit there and examine molecules as a traditional test would do. There are a lot of things that can be done with this technology. They're going to greatly improve directed therapies.

As you may know, some of the most advanced cancer therapies are monoclonal antibodies that target tumors specifically. Unfortunately, they target only a very small set of tumors. You have to choose your tumors carefully, but this allows you to get therapy against the tumors without all of the collateral effects. What does this have to do with nano? Is it just that we can make a better molecule? Molecules are just general-purpose things. Certainly, a general-purpose thing means that it's not particularly good for any special thing. If you design something that does a lot of different things—it's hard to do a lot of different things well. So what's happened in the drug industry is that they'll make a molecule and they'll move atoms around. They'll try to modify the bad effects down and get the good effects up, but it's still all one molecule that has to do five or six different things—that's very hard to do.

What's happening with these directed therapies in nanodrug delivery systems? This is an area I'm very actively working in. You effectively make a little nanovehicle which contains a collection of molecules, each of which is a specialist in its own area and can deliver treatment well. You're delivering a collection of molecules with very tiny implantable devices. You can see the tremendous progress we've made when we can put things small enough that they really don't affect the human body around them. We're already have implantable devices where you can swallow it—a little thing about the size of a pill and it has a TV camera in it and it literally goes down your gastro intentional tract and it takes pictures. It's a reality!

Wouldn't you like to have some things that would patrol around the rest of the body taking samples and tell you what is found? That can be done. We don't know how to do it for all cases quite yet, but it is certainly possible to do.

We're going to have new point-of-care devices where you can't even feel a pinprick because it's using a tiny laser beam you set on the surface. It samples the blood, does your complete blood chemistry, and sends it by telemetry over to the doctor's office. They analyze it and come back to tell you, "Hey, you look pretty good today. Everything's going fine." One of the biggest problems of health care is that you don't go to the doctor unless you're sick; unless you have big symptoms which, by definition, is already late in the diagnostic process. For good healthcare you should be monitoring your body all the time. In fact in the future, healthcare is going to be a very different thing than what it is right now. We will monitor health continuously with nanosensors and detect disease prior to traditional symptoms.

Future(istic) Possibilities

Given all these separate elements, I want to end by pulling together a number of things. This is my move from science back to science fiction. It's rumored that the head of NASA and the National Cancer Institute (NCI) just got together in a coffee shop in Washington one day in 1999 and decided to fund something really far out. They actually funded seven projects by NASA-held PIs, of which I was one, along with six National Cancer Institute PIs. Thirteen of us, who all still know each other, have gone on to do some great things. These were far-out projects at the time, but they've been fantastically successful.

We've gone from Da Vinci's *Vitruvian Man* to Isaac Asimov's *Fantastic Voyage*. Asimov imagined a ship that we shrunk to such a size that the vehicle could go around, patrol inside the body, and do things. Now, Isaac Asimov had a fantastic imagination, but he had one difficulty: he couldn't figure out how to put decision-making system into that vehicle without shrinking human beings down with it. That's why it's science fiction. But that's the only part of it that's science fiction. The rest of it we can do, though maybe in a slightly different form. What we're doing, what we're making, is multilayered nanoparticles that I call *programmable nanoparticles*. These particles will do a whole sequence of steps to do several different things.

However, when news of the NCI funding for this research first came out, it was very embarrassing. An article was written—and this is where the hype came in. I probably had a hundred radio stations and TV stations descended on me requesting interviews. I basically shoved them all away and said, "Go away, we haven't done anything yet. Come back in a few years and hold us to account." But we actually have done a good part of what we said we were going to do. It's taken a while, and it's taken a few detours, but the parts that I have pointed out here aren't Hollywood but real science.

Most of the time we do things in science we're not quite sure frankly whether it's going to work or not. That's a tough thing to say particularly if you're trying to get money for your research. You say, "If you give me this money maybe it will work, maybe it won't, I don't know." Nobody wants to hear that. A good thing about some of the work that I'm doing is that I'm certain it's going to work. And the reason I'm very sure is that it's already been done. Mother Nature has already done most of the things that I and others are trying to recreate synthetically. We just have to learn how to listen to Mother Nature. If you're interested in that you should check out this book called *Biomimicry* (Benyus 1998). One of the things it mentions is that nature has been doing nanotechnology for a long, long time. Nanotechnology is throughout nature, throughout your body. Your body has all kinds of little nanomachines inside right now doing all kinds of marvelous things every day. If we can imitate that, if we can study how nature pulled that off, I think our hypothetical technologies and devices can become realities.

For instance, consider what is called in modern terminology "twenty-four-seven surveillance" of your body. You never know whether you're sick or not unless you get a big enough symptom, a bad enough symptom. You go to the doctor since you have a fever and a lump. By the time you have a lump, that's millions and millions of tumor cells. That's not a good time to find out. So, conventional medicine is reactive to tissue-level symptoms. We have to have a symptom or we don't even go to the doctor. So, what happens when you find out that you've had cancer for the last five years and you didn't even know it? And if you'd treated it five years ago the prognosis might have been very good, whereas now it may be fatal.

So we're going to treat things prior to traditional symptoms. I spent twenty-seven years of my life as a professor at medical schools. When I first started giving some of these public talks on this topic, I would say: "You know what, put away your stethoscopes. You're going to have to start thinking pre-symptomatically about things. You're going to have to look at medicine at the molecular level inside individual human beings and stop doing things traditionally." Now as a doctor you say, "This has a 90 percent probability that it's going to fix you right up. It's got a 9 percent probability that it's not going to do you any good and a 1 percent probability that it's going to kill you." Does that give you or your patients a lot of confidence? It doesn't give me a lot of confidence! I was in one of those 1 percent once and almost died due to one of those exceptional circumstances. It was not a great experience! The whole concept of when you actually know whether you're sick or not is going to change. First, because we can diagnose so much earlier, we'll be sitting there watching what's happening at the molecular level.

Second, we will be able to refine the art of medicine. I'm not going to say that physicians and their knowledge and experience aren't going to be important: it's a wonderful thing to have those physicians. But you know what? Two-thirds of the world has no doctors. In fact the most rural, poorest parts of the United States have almost no doctors because becoming a doctor is a very expensive thing. It's an expense of a person's time, and also a financial expense: you can't make enough money to even pay for the education by going out and practicing medicine in rural America, much less in third world countries. So most of the world doesn't have any diagnostics and limited medicine. But what if we could mass manufacture very small nanomedical devices? Those could be at least the first line of diagnosis and defense to detect viral infections and disease early. This is another of my dreams. Probably it won't happen in my scientific career or lifetime, but it is the future of nanomedicine and can change the world.

Third, I mentioned the "fantastic voyage of the nanosurgeons," and I joked by saying: "Well that's the only form of surgery I'm allowed to practice. And that's probably a good thing. I'm allowed to perform surgery only on molecules, not on people or their tissues." But what if this voyage was not fantastic but reality? That's what we're doing: we're developing ways to go in and reshape molecules. Cells are diseased because they have the wrong chemistry. If we could redirect the chemistry, we could possibly change a diseased cell, if not back to a normal cell, then to at least a less diseased and less dangerous one. This is not a pipe dream. Could we make a disease cell into a less dangerous diseased cell? I think the answer is going to be yes.

When you see the word nanomedicine, first of all, the nano- is there not just because of the size—a billion times smaller than I am. That's just one of the reasons. But, it's also there because of the approach. I've talked about sculpting big things into smaller things. But you know now that we can also make larger objects from the ground up, sort of a bottom-up approach of nanotechnology. We need special tools—nanotools and other smart things. Whenever you work with anything, you have to have tools the size you're working with. Recall Richard Feynman's big fingers: we've got to make even skinnier fingers so we can lift atoms up, move them around, examine them, and know what we're working with. So inventing and making tools that are smaller and smaller: this is a bottom-up approach.

Instead of thinking about things as tissues and organs as we do right know when we cut them out, burn them by radiation, or poison them out with chemotherapy, we want to think of them as groups of cells—we really want to do individual targeting. We want to fix the cells without destroying the organ. A primary problem is that doctors have, by trying to take out diseased cells by traditional therapies, destroyed the function of the organ and patients have died.

When I first got into medicine, I was horrified. At MIT they really taught us to pay attention to feedback loops. You never invented something that didn't have a feedback loop. A system without a feedback loop was a runaway system, when things got out of control. Guess what? Medicine has never had feedback loops in it. We hit people with medicines. You take a dose of something and you don't take another dose for four hours. You spike the effect up, it decays, and then you spike it up again and it decays, and you're supposed to have some mythical average in all of that. Now however we have time-release capsules in drug technology, which evens that pattern out a little bit more, but the basic idea is you're almost never responding directly to the disease at the best therapeutic dose. You may distribute the chemical in some way, but you don't even know what you did. So what I'm saying is these new nanomedical systems are going to literally have sensors, switch sensors, and you can make these out of molecules that actually go out and they say, "What happened to what I was trying to deal with. Is it still there? Is it still a problem?" This is yet another thing that nanotechnologies are going to allow us to do.

My lab designed a nanoparticle that can accept a core. One of the things we're doing is we're taking ferric oxide core, an iron particle, as a core. We're trying to build upon those kinds of structures, developing drug delivery systems from the inside out. The next layer out has therapeutic drugs and genes. We talk targeting, but this nanoparticle system is one-millionth the volume of a blood cell in your body. Imagine if it comes up to your blood cell and says "I'm here," targets the cell, and then drops its payload off at the front door. How many drugs can you dilute a million-fold and still have them do anything? The answer is very few. So you not only have to get it to the right cell, but also to the right region of the cell. We call this *intracellular targeting*. We've done this with molecules that will target to a particular three-dimensional local in a cell. Then we move to cell targeting. Finding rare cells is like the needle in a haystack. Putting nanoparticles in the body to find these very rare cells is difficult. By definition, they're rare.

Nowadays, you are putting massive amounts of drugs and chemicals in your body. When you ingest or inject these drugs, basically they go everywhere in your body. Which is why they say, "This was a treatment for your headache but, by the way, this may cause liver damage." What we're trying to do is make something that doesn't go to the liver if it is supposed to deal with your headache. You can actually build in the chemistry of this, and this is where Isaac Asimov didn't think about it correctly. If he had thought about how you make something that has decision-making power, he would have made sure that the chemistry happens in a very particular order. Once you've done that, you've effectively created the equivalent of a chemical computer program that has to go step A, B, C, D. If it can't get past step B, then steps C and D are not going to happen. So if it goes to the wrong place, if it targets something, it goes inside and says: "Now I'm going to look for my target inside the cell, and I can't find it." Then it's the equivalent of making the decision: "Oops, I went to the wrong cell, so I shouldn't do the rest of my programming. I shouldn't deliver any message." Can we do that? We absolutely can. We can put in error-checking and all kinds of things. As it goes to the cell, we have some chemistry in there that will cause the first layer to peel off. Then, it goes through the cell membrane and targets the next layer. There is a lot of chemistry in this, a lot of magic in the details so I won't say that all of these steps have been done, but we've done most of this. And the next step is therapeutic genes. Genes have gotten a bad name because people were not very careful with what they we're doing. But a gene is just another form of drug. In fact, the most natural thing in your body is DNA and RNA. So, the most natural drug that you can make is made out of DNA or RNA. So we're making a molecule, just like any other drug, that's made out of DNA or RNA. The goal is to make these multistep targeting systems biodegradable.

Now, that sounds fantastic. But I didn't think of that idea in the first place. It was done millions of years ago by Mother Nature. It happens every day in your bodies and other places. We know well one of nature's nanoparticles that, when it's doing what it wants to do, is considered bad—the virus. But it turns out these nanoparticles that we are making are the same size as a virus, except they're not viral. They're completely artificial made out of safe, biodegradable substances.

So here's the idea: I call it the *nanofactory template*. All we need to do is have a few molecules that go in and create a template. Upstream, we have what's called a *promoter*, which is the on/off switch to the gene. Upstream from the promoter we put a *bimolecular sensor switch*. That's a molecule that is easy to build. When it finds the target inside the cell, it binds to the molecular biosensor, turns on the switch, and starts producing this therapeutic gene. It's treating the target; it's blocking and degrading the bad molecule inside your cell that's making your cell diseased. When those molecules run out, guess what, there's none of them left to land on the gene you make. This is a gene that automatically turns off unless it sees its target. So it doesn't do anything. It turns on and off. You could conceivably do this with chronic disease, which literally turns on and off as the disease recurred. That's quite a bit harder to do functionally but it's theoretically possible. It manufactures the therapeutic gene out of the raw materials inside the cell—in the body. It just takes the basic nucleotides of DNA or RNA, depending on how you made it, and reconstitutes those. This is exactly what viruses come in and do in your body all the time. They steal your DNA and RNA and they reformulate that genetic material to make more copies of themselves which ultimately do damage. The problem is they do it for their purposes, not your purposes. So let's use those materials and then actually let a template manufacture a drug.

If I'd told you this in the beginning of this essay, you would have said, "This is science fiction." But everything I've told you here is very doable. In fact, all the pieces have been made. What has not yet been done is to make good, integrated systems that all work well together. We know it's easy to make each part separately, but to make them all work at the right time and in the right order is a whole lot harder. I think nanotechnology is going to have a tremendous effect on that. Once we can do something better and make it available to everyone, there starts to be a moral imperative to actually deliver nanomedicine to everyone and not create a "nanomedicine divide." There are many cases of this happening in medicine whereby once it was possible to provide a treatment (that was previously impossible or non-existent) it became unethical to not provide that treatment. I think that many of the benefits of nanomedicine *must* be made available once they are possible and practical. As always, political and economic factors will enter into that equation, but we should not deny the benefits of nanomedicine to all of our people and create an even bigger "dual track" of medicine than we have now. Many years ago both rich and poor really did not receive very different standards of medicine because medicine in its infancy could not provide much real therapeutic effect. But that is no longer the case and will be even less so once the full benefits of nanomedicine are appreciated.

Nanomedicine is where some of the hype and debate, the science and futuristic science fiction, intersect. Some of it is may still be purely fantastic science fiction. But, there's also science fiction which is just science that hasn't happened yet. We have to be aware that we're living with a lot of yesterday's science fiction that is just a lot of normal, everyday stuff to us now. So the time to worry about the bioethical dimensions of nanotechnology is now!

Notes

1. "There's plenty of room at the bottom" is the title of Feynman's lecture at the American Physical Society meeting at Caltech on December 29, 1959. A reprint of the talk was published in 1992 in the *Journal of Microelectromechanical Systems*, 1 (1): 60–66.
2. Rick Smalley passed away in October of 2005. The person you'll often hear of now, who has also taken Drexler to task, is George Whitehead of Harvard.
3. Open access e-copy (html) available on http://e-drexler.com.

References and Recommended Reading

Altmann, Jurgen, and Mark A. Gubrud. 2004. "Military, Arms Control, and Security Aspects of Nanotechnology." In *Discovering the Nanoscale*, edited by D. Baird, A. Nordmann, and J. Schummer. Amsterdam: IOS Press.

Baum, Rundy. 2003. "Drexler and Smalley Make the Case For and Against Molecular Assemblers." *Chemical and Engineering News* 81 (48): 37–42.

Benyus, Janine M. 1998. *Biomimicry—Innovation Inspired by Nature*. New York: HarperCollins.

Berube, David M. 2004. "The Rhetoric of Nanotechnology." In *Discovering the Nanoscale*, edited by D. Baird, A. Nordmann, and J. Schummer. Amsterdam: IOS Press.

Berube, David M. 2005. *Nano-Hype*. Amherst, NY: Prometheus Books.

Drexler, Eric. 1986. *Engines of Creation: The Coming Era of Nanotechnology*. New York: Anchor.

Drexler, Eric, and Chris Peterson. 1991. "Unbounding the Future: the Nanotechnology Revolution." *Foresight.org*. http://www.foresight.org/UTF/Unbound_LBW/index.html.

Feynman, Richard. [1959] 1992. "There's Plenty of Room at the Bottom." *Journal of Microelectromechanical Systems* 1 (1): 60–66. doi:10.1109/84.128057.

Freedonia Group. 2005. *US Nanotechnology Health Care Product Demand to Reach $6.5 Billion in 2009*. http://www.nanotech-now.com/news.cgi?story_id=09445.

Johnson, Ann. 2004. "The End of Pure Science: Science Policy from Bayh-Dole to the NNI." In *Discovering the Nanoscale*, edited by D. Baird, A. Nordmann, and J. Schummer. Amsterdam: IOS Press.

Minsky, Marvin. 1986a. Foreword. In *Engines of Creation: The Coming Era of Nanotechnology*, edited by Eric Drexler. New York: Anchor.

Minsky, Marvin. 1986b. *The Society of Mind*. New York: Simon and Schuster.

MIT. 2012. "MIT Institute for Soldier Nanotechnologies—Research." *Massachusetts Institute of Technology*. http://web.mit.edu/isn/research/index.html.

Moor, James, and John Weckert. 2004. "Nanoethics: Assessing the Nanoscale from an Ethical Point of View." In *Discovering the Nanoscale*, edited by D. Baird, A. Nordmann, and J. Schummer. Amsterdam: IOS Press.

National Nanotechnology Initiative (NNI). 2007. "Nanotechnology and Life Cycle Assessment: A Systems Approach to Nanotechnology and the Environment." *The Project on Emerging Technologies*. http://www.nanotechproject.org/file_download/files/NanoLCA_3.07.pdf.

NNI. 2012. "Nano." *National Nanotechnology Initiative*. www.nano.gov.

Roberts, Jody A. 2004. "Deciding the Future of Nanotechnologies: Legal Perspectives on Issues of Democracy and Technology." In *Discovering the Nanoscale*, edited by D. Baird, A. Nordmann, and J. Schummer. Amsterdam: IOS Press.

Robinson, Wade L. 2004. "Nano-Ethics." In *Discovering the Nanoscale*, edited by D. Baird, A. Nordmann, and J. Schummer. Amsterdam: IOS Press.

Roco, Mihail C., and William Sims Bainbridge, eds. 2001. *Societal Implications of Nanoscience and Nanotechnology*. NY: Springer.

Sandler, Ronald. 2009. *Nanotechnology: The Social and Ethical Issues*. Washington, D.C.: Project on Emerging Technologies. http://www.nanotechproject.org/process/assets/files/7060/nano_pen16_final.pdf.

Saxton, Jim. 2007. *Nanotechnology: The Future is Coming Sooner than You Think*. Washington, D.C.: Joint Economic Committee. http://www.coalitionoftheobvious.com/nanotechnology_03-22-07.pdf.

Schuler, Emmanuelle. 2004. "Perception of Risks and Nanotechnology." In *Discovering the Nanoscale*, edited by D. Baird, A. Nordmann, and J. Schummer. Amsterdam: IOS Press.

UNESCO. 2006. *The Ethics and Politics of Nanotechnology*. Paris: United Nations Educational, Scientific and Cultural Organization. http://unesdoc.unesco.org/images/0014/001459/145951e.pdf.

Woodrow Wilson International Center for Scholars. 2013. "Project on Emerging Nanotechnologies." *The Project on Emerging Technologies*. http://www.nanotechproject.org/index.php?id=44.

Biographical Sketch

James Leary, http://web.ics.purdue.edu/~jfleary/, is an endowed professor of nanomedicine, professor of basic medical sciences, and professor of biomedical engineering at Purdue University. He is a member of several research centers at Purdue, including Bindley Bioscience Center, Birck Nanotechnology Center, and the Oncological Sciences Center, and was elected in 2007 to the American Institute for Medical and Biological Engineering. He received his PhD from Penn State University and completed a postdoctoral fellowship with Los Alamos National Laboratory. He holds at least nine US patents and has published over 135 papers in the field of science and nanotechnology.

Ethical Issues in Constructing and Using Biobanks (2008)

ERIC M. MESLIN

There is a wide range of interests in biobanking. Some people are interested in the topic because they came across the term in their research. Others are interested because they have seen descriptions or depictions of biobanks in the media. Those who have been thinking about this issue or who have contemplated the ethics of donating parts of their bodies, whether that is via an organ donor card in a wallet, a blood test, or a biopsy that was done at a hospital—maybe a suspicious mole or something worse removed—may have wondered what happened to those pieces of themselves. In what follows, I am going to tell you a little about where some of the challenges are in constructing what we're calling biobanks, or biorepositories.

Some Definitions and Some Data

I might as well start with some definitions. Imagine a bunch of stuff in a fridge: there is nothing more complicated to biobanks than that. There are lots of technical definitions around the world, but I will use this one as a working definition: "The term *biobank* refers to collections of biological material, for example blood or tissue samples. These samples . . . can often be linked with information about the donor's health status" (Codex 2010). Generally a biobank refers to a collection of biological material, like blood or tissue samples (and I'll say more about that in a second), which are collected and stored in a fridge, sometimes under very cold temperatures. But that's all it is. What's cool about these things is that we can do things with them—not only with the samples they hold but also with the information contained in that material. So that's our working definition for what follows here.

What are these biobanks and where did they come from? There's been lots of work done on this issue of biobanking, including by the National Bioethics Advisory Commission (NBAC), which I ran for almost four years. The collections come in a variety of forms, using a variety of sources for samples. Some banks are used for longitudinal research studies and consist of simple blood or urine specimens; some banks have more specialized contents such as banks used for newborn screening tests. Most parents were probably like me: you almost fell on the floor when your first child was born and they take her, turn her over, and stick a pin in her heel in order to test for PKU (a potentially lethal disease easily treated by diet). I know I was floored when my daughter's heel stick was done. Public health departments have thousands and thousands of these so-called "Guthrie cards," with little blood spots on them.

On April 20, 2007 Indiana Governor Mitch Daniels signed a bill that established the first public umbilical cord bank in Indiana (Indiana Code, Title 12, Article 21, Chapter 1) allowing women who give birth to have the option of donating the umbilical cord and the blood contained in the cord to a bank to be later used for both treatment and research. This is a wonderful and non-controversial example of a public biobank. Usually the cord and related material is thrown away after birth and yet now we know that it's incredibly valuable as a very rich source of stem cells.

One of my colleagues and staff members at NBAC was an American Association for the Advancement of Science (AAAS) fellow named Dr. Elisa Eiseman, who worked with us under contract from the RAND Corporation. She wrote the definitive background report on this topic for the Commission and as well as other key literature (Eiseman & Haga 1999). Dr. Eiseman's work may be a little out of date now but is still very instructive. In 1998 there were about two hundred and eighty two million specimens from these various sources that were stored in a variety of biobanks around the country. The number is actually closer to about three hundred and fifty million specimens. In fact we don't know what the definitive number is because when she first did the study for NBAC, Dr. Eiseman included data only from centers and banks that voluntarily reported their numbers. There are lots of other banks out there, and in fact there are many more than we don't know of for proprietary reasons that are housed, for example, at pharmaceutical companies or biotech companies.

They say, "I'm not interested in telling you what I have or why I have it." They're entitled to say that. One of the largest is the Armed Forces Pathology Institute lab at Fort Detrick, Maryland. They have loads of samples, many more now than before given the requirement to obtain samples from all servicemen and women before they go to war to aid in the identification of remains should that be necessary.

In fact we are so awash in biological material that we don't really have any idea how much there is in this country, let alone in other countries. Any university would be hard pressed to tell you exactly how many refrigerators they have and what's contained in them. That's not because they're unethical and it's not because they're acting inappropriately: they're just not required to keep track of every single cell in every single test tube in every single fridge/freezer. There's no federal law that says you have to tell me how many biosamples you have in your refrigerators. I'll show you why I think that's important soon.

The fact that we have all these biosamples and have had them for a long time is not terribly surprising. We've been using specimens and samples for years. In fact we've been using them for as long as there have been dead bodies and inquiring people, who have taken pieces of them and done things to them: smelled them, tasted them, examined them. The whole field of pathology began by the examination of human specimens to learn about how the body functioned. Surgeons have been doing this for several hundred years, heading to some very important medical benefits. But for the use of already stored samples we would not have been able to do a long list of important medical tests and diagnoses.

The pap smear would not have been invented by Dr. Papanikolaou whose name graces that particular test. It was the study of precancerous lesion samples from women that lead him to start to develop and perfect what is now a very common and useful diagnostic test. We know about lung cancer much more as a result of the autopsies performed on smokers with emphysema, whose lung tissue was stored under optimal conditions. Everyone has been talking about the next great pandemic influenza outbreak, and lot of people are tracing current flu outbreaks back to the great flu epidemic of 1918. Researchers from around the world sequenced the SARS virus in about six months because they were able to collect those samples quickly and efficiently. They were able to determine what virus it was. We've been doing a lot of that same work for a long time, on HIV.

Those are just examples from two fields like cancer and infectious disease. We could also discuss the use of samples by the device industry, which has used and relied on larger samples like body parts, hip joints, and long bones. They too have collected and stored and carried out research not only diseases of those particular body parts but on how to build orthopedic devices and the like. But it is the biomedical research uses of human biological materials that I mostly want to focus on here.

Building Biobanks

Among the most common and easy to understand uses of biological material and methods of building biobanks have been the many large longitudinal studies. In several cases we have been following individuals, asking them to come in for regular medical visits and to provide a blood sample so that we could find out what the causes of—or at least the risk factors associated with—various diseases over a long period of time. These include large groups like the Nurses' Health Study, the Nun Study of Malta, and many others. In the US the National Institutes of Health (NIH) have sponsored many of these studies. We've been doing this a long time.

I don't know if anyone has been to downtown Framingham, Massachusetts, but it's a pretty cool place; so cool in fact that as early as 1947 it was home to one of the world's most important and interesting longitudinal studies looking at the causes of heart disease. Researchers went to essentially everybody in this town and said, "We'd like to research you, to keep in touch with you, for the rest of your life. Would that be ok? We'd like to see when and if you die of diseases like heart attacks or stroke, the major cardiovascular diseases, and see if we can figure out what's causing them. Maybe it's your diet, maybe it's your lifestyle." In 1947 we didn't know a much about the human genome, but we knew a little about people's behavior, what they were eating. We knew a little about cholesterol then, not as much as we know now, and the like. This is a study that doesn't have an end (Framingham Heart Study [1948] 2011). In most medical research, the typical model of biomedical research is that it has a beginning and middle and an end in usual the three to five or maybe even ten year period.

The biobanking facilities that I have mentioned here so far are by no means the only such places on the planet. Let me give you some examples of where else large biobanking studies are being done. The group called DeCode Genetics in Iceland got permission from the Icelandic parliament to collect samples of everybody. Interestingly, they then licensed the information collected from those people to a private company so they could start mining the information. Iceland has what's called in the genetics world a very strong *founder effect*, which means everyone seems to stay in Iceland, and there's not a lot of intermixing with genes from elsewhere. There are several places around the world like that: Newfoundland and some parts of Quebec in Canada, Papua New Guinea, too. There are little pockets of—I don't want to say "pure" since that's an ethically charged term—people without a lot of "misspellings" of the genome caused by folks from other places. The United Kingdom too has now collected more than five hundred thousand samples from its people in a project called UK Biobank (http://www.ukbiobank.ac.uk/). CARTaGENE is a Canadian biobanking project group (http://www.cartagene.qc.ca/). The Swedish national biobank program is another (http://www.biobanks.se/). Generation Scotland is the name of the Scottish biobank that's starting now (http://www.generationscotland.org/) and you can find banks in Australia, New Zealand, Switzerland and a growing number of South American countries.

There are also biobanking developments going on in west and east Africa, including one in Kenya which we are involved in through a program supported by the IU School of Medicine. Within the next five years, this will be the biological equivalent of keeping up with Joneses: if you don't have a biobank, you're nowhere. It's fascinating to see that in so many culturally diverse places, collecting parts of people and storing them in a fridge for other people to use is not a common thing in the course of human affairs but something everyone is doing.

Three Factors Ramping Up Interest

This quick growth went relatively unnoticed until at least three things happened that I think have ramped up the interest in both the use of this material and, in many ways, the ethical challenges associated with this field of biobanking.

First, I would simply call it the "accelerant effect" of the human genome project. The Human Genome Project (HGP) was the effort to map genes—find their location on the human chromosomes, and then to sequence them by determining the order of the letters of the genetic alphabet that made up each gene. Through a combination of public and private funding, scientists figured out the approximate number of genes and their approximate "spelling" using the three billion letters of the genetic alphabet, and the data was offered up in papers in *Nature*, volume 409, issue 6822 and in *Science*, volume 291, issue 5507.

Well clearly that was interesting, but that wasn't really where all the action was. Mapping the genome didn't tell you anything about why certain people were going to get heart disease or why others were going to get diabetes. We didn't know what those genes did. It was like having a copy of *War and Peace* without any punctuation, page numbers, chapter headings, or anything to set off the letters in all those sentences. Now multiply the book by twenty-seven. The human genome's three billion letters is like having twenty-seven copies of War and Peace without the punctuation, page numbers, or anything like that. And the fact that War and Peace was originally written in Russian is not so far off the mark for that analogy, because the genome was written in a language that we didn't fully understand. However as technology progressed, this sequencing increased the value of the genome and of all the genetic material kept in fridges forever more.

The second thing that contributed to our interest in the construction of these biobanks was the dramatic rise in the informatics world. Medical informatics, bio-informatics, the world of information technology itself—the very rough picture of converting hard copy files on a shelf into digital information—may seem quaint these days. But it was a phenomenally important advancement in the world of biological science because we now have genotypic information that we could start to correlate with *phenotypic* information, the actual facts of an individual's appearance or medical history. This has led to an equally large information explosion. I'm sure many working in the science of genetics think that's where all the cool work is being done. Well in many ways, bioinformatics may be where the action is. It's not very interesting to merely find out what the genetic alphabet is. We know that, we've got it, we've sequenced the genome of many, many organisms now. It's a run of the mill technology. The exciting part is that the computer technology that now exists can do fantastically fast and complex calculations and predictions of how genes and genetic mutations will influence disease. They will also help design better drugs by using genetic information to determine whether a person will respond to a medicine or not.

The take-home message here is that the value is in the information contained in the material. Having a lump of you is interesting but absolutely irrelevant. Science moves in very strange and mysterious ways—two steps forward, one step back, and another one to the side. It doesn't make promises. James Watson made the audacious claim and prediction in 1989 that we would map and sequence the genome in ten years for three billion dollars. That is the worst possible thing that anyone could ever say to the US. Congress, because not only is it a little bit immodest to make such a claim, but also because we didn't yet even have the tools to map and sequence. But a very strange thing happened on the way to the mapping and sequencing of the genome. Science started to build the tools. We started to drive down the costs of sequencing: work that originally cost two, three, five, or even ten dollars a base pair soon cost twenty cents a base pair, and the cost is even lower now. Informatics scientists started to work closely with genetic scientists. The genome became a playground for mathematicians, geneticists and computer scientists.

Besides the accelerant effect of the genome project generally and the rise of sophistication of health and other informatics technologies, the third of the three factors that I think are contributing to our discussion about biobanking is the drive to commercialize, which really drives a lot of research and innovation in this country and others. It's no longer the simple situation that there is scientific value in the little pieces of you. There is actually financial value of some kind if we can only extract that information. Commercial products can be any number of things. They can be drugs or tests or devices. We have a very strong incentive system in this country. We want to create an environment where there is an incentive to invest money—public, private, philanthropic— because that investment will bring people to the table and build really cool buildings that we can fill up with really smart people. And much like those rooms full of computers that Craig Venter had at his private company Celera, we too can fill buildings with really smart people working together. Well, you need money to do that. You need money to build research parks at universities, to buy the technology and pay for graduate students, and to pay to keep the lights on. The way that we do that in this country is to have an unofficial agreement that universities are often the place where we do basic research, and the private sector is where we commercialize or apply that research. This was the historic agreement started after the Second World War. That agreement has now changed—it hasn't so much been ripped up, but rather been modified significantly. Parts have been erased, typed over, whited out, blocked and replaced, and now you can find universities that are every bit as entrepreneurial as the biggest and best pharmaceutical companies and biotech companies who are so solely run on venture capital funds that their research is as "pie in the sky" as any university's. Directors have raised funds and are doing basic work at the level of molecule.

So the world is changing quite a bit as a function of this translational activity, trying to get more knowledge, and more data, that we can hopefully turn into health benefits. A good example here is blood samples. Take, for example, a bunch of patients who are going to have cardiac catheterization, a procedure in which a little probe is snaked through cardiac arteries to clear them out. Blood can be taken and used for various types of genomic research.

Meanwhile we also want to do things like mine large bioinformatics repositories, the electronic medical records or databases. We also do surveys on people, find out what their medical histories and health behaviors are. If we put all those three things together—the piece of you, the health information about you from your medical record (you have heart disease or your parents had heart disease), and then details about your life style, we are going to learn an awful lot more about you. That's the model of how you use a biobank.

So far I've explained the science and technology in very quick and rough terms, but the principal ethical hurdle to achieving this success is the willingness of people to *consent*, to say, "Yes, you can take a blood sample when I come into the hospital," or "You're doing that biopsy for my little mole, yeah you can use that in research."

Informed Consent

We know it wasn't always the case that people were asked for their permission to use biological materials. Surgical consent forms from the 1960s and '70s often included at the very bottom of the form, in little tiny font, a statement that informed the patient that any material obtained in surgery might also be used separately for research. So an awful lot of research was done on that leftover waste without the informed consent of patients.

Informed consent for medical research is an extremely important and federally required activity. You cannot do medical research on human subjects at any university in the country unless you have their prior, informed consent, and a committee called an Institutional Review Board has approved the research. It's a big deal. It's not an optional ethics thing. If you don't do it and the FDA finds out about, they can shut the university's research programs down. Don't believe it? They've done it forty times in the last ten years. Forty times. They shut down Duke (Weiss 1999). They shut down Johns Hopkins (Steinbrook 2002). You may not have heard about this because they don't like to publicize the fact that the federal government turned off the spigot of funding. But that's how important informed consent is—just that one issue.

Unlike typical medical research, for example, a drug study, biobank research often does not have a specific starting and ending point. The longitudinal studies I mentioned above are examples of that. In fact a biobank study can go on long after a person has died. How do you consent to something that will happen after you're dead? Is that even possible? If you're not willing to stretch your mind that far, all you have to do is go to an area like degenerative Alzheimer's, a very important field of study. How do you seek informed consent from someone to do research on Alzheimer's or when you're really interested in whether the drug you have can reverse end stage Alzheimer's? The very moment that you want to try a new drug on somebody, they lack the capacity to consent to participate in that study. Maybe you don't do research on people who can't consent. That would solve the ethics problem. It would also mean we couldn't do research on children because children are not legally allowed to consent as long as they are minors. We couldn't do research on trauma victims. We couldn't do research on people in emergency rooms to find the best way of resuscitating a patient. Before we sort of throw out the issue and say, "Well let's just have consent for the things that are happening to me right now," we need to consider what else we might be throwing out.

On this issue the world of biobanking tends to be conceptually divided into at least two very broad problems. First, we have to think about the stuff that's been in the fridge for a long time for which there may not have been formal informed consent. Can we use that sort of stuff? You certainly could use it if it was collected without identifiers, that is, collected anonymously. We did that in the early 1980s with HIV zero-prevalence studies. We wanted to find out how many people in a community were HIV positive, but was a very controversial issue to find out that someone was HIV positive at that time. Individuals might be discriminated against, stigmatized, or denied health care. The only way we could do those studies was to take blood samples from everyone, no names, no identifiers, no addresses, nothing. We just want to see, in all the test tubes that we collected from this area, how many of them are HIV positive and how many of them are not. That would be a way to do research on the stuff already in the fridge, right? Or you could probably strip all those identifiers and still do all the research. Federal regulations for medical research say if you're doing research on anonymous samples, unidentifiable, then you don't need informed consent. How would you get informed consent since you don't know who they are in the first place?

There are some purists who say, "What we really want is to know who the person is, whose sample is in the fridge because we want to match it with their health records" and thus do all the things that I told you about earlier. You don't find much valuable scientific information from anonymous samples. So, what should we do? Should we re-contact everybody and get consent from them? Well that's a very labor intensive, very expensive proposition, but that would be a possibility. Or we could figure out a way to just waive consent and say "the statute of limitations has expired, so we're just not going to worry about it."

So we're still trying to figure this out. Such ethical problems with research on currently stored samples haven't been solved yet. On the one hand, there is valuable information in those samples—scientifically valuable information. On the other hand, there's a moral principle at stake, a principle of personal autonomy, or of human dignity—a personal respect for the decision making of others. This is a pretty significant issue. If you're going to treat it casually or in a cavalier way, you're risking an awful lot more.

Let's assume that we've solved this dilemma and figured out how to deal with the currently stored stuff in the fridge. The next big issue, if you think back on the global scope I told you about, is that we want to start collecting new samples now where there's no worry about informed consent. We'll just ask everybody, "Would you be willing to consent?" in exactly the way we ask you, "Do you want to sign your driver's license as an organ donor?" You make an affirmative consent, "Yes, I do," and they put a little heart on your license, or "No, thank you I don't want to," and that's ok too. No one is going to force you to do it. We have an opt-in system here in the US. But what would an opt-out system look like—where everyone was presumed to consent to participate in a biobank unless they specifically declined? Opt-out would almost certainly result in more specimens being collected for research.

Opt-out policies also have challenges of their own; the most obvious is that they don't give patients the opportunity to decide what type of research they want to participate in. The further away from the activity for which your consent is provided, the less informed it is. It also runs the risk of the public losing their trust and confidence in science. Some of these ethical issues are informed by empirical research about the public's willingness to donate. We've done a number of studies at the IU Center for Bioethics on this problem, as have others around the world. One of our studies, conducted at the IU Cancer Center, found that 60–85 percent of subjects agree that stored tissue can be used in unspecified future research (Helft, Champion, Eckles, Johnson, Meslin 2007). We also did a study with women and pregnant mothers (Haas, Renbarger, Meslin, Drabiak, Flockhart 2008) that produced some similar support for future research.

Observations: Regulation, Privacy, and Commercialization

Let me conclude with a couple of final observations. First, biobanking is not happening in a vacuum. It's happening within the context of a legal and regulatory environment in this country represented by the three branches of the government: legislative, executive, and judicial. And it's happening at both the federal and state level. We haven't seen a lot of action at the US Supreme Court yet. But we're starting to see the States weigh in. We also know that our current system for regulating human subjects research needs updating to handle genetics studies. This has created what my colleague Barb Evans and I call the *harmonization problem* (Evans and Meslin 2006): different rules are being used by different branches of government.

We know, based on a state-by-state analysis that Girod and Drabiak did (2008), that the rules for conducting biobank research vary greatly depending on whether you are in, say, Indiana, New Mexico, or Alaska.

Second, privacy concerns issues will continue to be important impediments to biobank research. Recall the very stylized picture from the front page of the New York Times a few years ago (Baker 2008), in which genetic banks were portrayed as Big Brother. And yet we also know that the more information a researcher learns about a person (reducing privacy in the process) the more accurate the research and the more applicable the findings to others who share the same genetic background. This presents a tough balance.

Finally, we come full circle back to commercialization. I told you that commercialization is one of the drivers behind the use of biobanks. It drives our ability to bring in money so that we can do more research. The main payoff, of course, is patient care. But we should also be mindful of the more obvious payoff—we want to get something out of it.

Many people are very worried about commercialization, not because they are worried about commodification of the body as a moral wrong based on the precept that the body is a temple and shouldn't be paid for. We think prostitution and slavery are moral wrongs because in some way it treats the body as some form of subhuman chattel. Who sponsors the research also does not upset people—whether it's federal sponsorship, NIH, university sponsorship, or drug company sponsorship. The most prevalent objection to the commodification issue is "Where's my cut of the action? I don't mind giving biomaterial to these banks, I'm more than happy to do that. But I'd like something in return, thank you very much. Where's my piece of the patent?" So we're still learning about the commercialization issue.

Indeed in some ways we're still learning about donation in general. We're learning things from the donation systems for blood and for organs, as well as from the world of philanthropy (Meslin, Rooney, Wolf 2008). But we still have a long way to go to understand the entire ethical situation surrounding biobanking.

References and Recommended Reading

Baker, Al. 2008. "Genetic Bank Raises Issues of Practicality and Privacy." *The New York Times.* January 19, http://www.nytimes.com/2008/01/19/nyregion/19dna.html?_r=0.

Codex. 2010. "Biobanks." *Codex: Rules and Guidelines for Research.* http://codex.vr.se/en/manniska4.shtml.

Framingham Heart Study. [1948] 2011. "Framingham Heart Study." *Framington Heart Study.* http://www.framinghamheartstudy.org/.

Haas, David M., Jamie L. Renbarger, Eric M. Meslin, Katherine Drabiak, and David Flockhart. 2008. "Patient Attitudes Toward Genotyping in an Urban Women's Health Clinic." *Obstetrics and Gynecology* 112 (5): 1023–28. doi:10.1097/AOG.0b013e318187e77f.

Helft, Paul R., Victoria L. Champion, Rachael Eckles, Cynthia S. Johnson, and Eric M. Meslin. 2007. "Cancer Patients' Attitudes Toward Future Research Uses of Stored Human Biological Materials." *Journal of Empirical Research on Human Research Ethics* 2 (3): 15–22. doi:10.1525/jer.2007.2.3.15.

Eiseman, Elisa, Gabrielle Bloom, Jennifer Brower, Noreen Clancy, and Stuart Olmstead. 2003. *Case Studies of Existing Human Tissue Repositories: "Best Practices" for a Biospecimen Resource for the Genomic and Proteomic Era.* New York: Rand Publishing. http://www.rand.org/pubs/monographs/2004/RAND_MG120.pdf.

Eiseman, Elisa, and Susanne Haga. 1999. *Handbook of Human Tissue Sources.* New York: Rand Publishing.

Evans, Barbara J., and Eric M. Meslin. 2006. "Encouraging Translational Research Through Harmonization of FDA and Common Rule Informed Consent Requirements for Research with Banked Specimens." *Journal of Legal Medicine* 27 (2): 119–66. doi:10.1080/01947640600716366.

Girod, Jennifer, and Katherine. Drabiak. 2008. "A Proposal for Comprehensive Biobank Research Laws to Promote Translational Medicine in Indiana." *Indiana Health Law Review* 5 (2): 218–50.

Meslin, Eric M., Patrick M. Rooney, and James G. Wolf. 2008. "Health-Related Philanthropy: Toward Understanding the Relationship Between the Donation of the Body (and Its Parts) and Traditional Forms of Philanthropic Giving." *Nonprofit and Voluntary Sector Quarterly* 37 (1 suppl): 44. doi:10.1177/0899764007310531.

Steinbrook, Robert. 2002. "Protecting Research Subjects—The Crisis at Johns Hopkins." *New England Journal of Medicine* 346 (9): 716–20. doi:10.1056/NEJM200202283460924.

Weiss, Rick. 1999. "U.S. Halts Human Research at Duke." *The Washington Post,* May 12, A1.

Biographical Sketch

Eric M. Meslin, http://bioethics.iu.edu/people/meslin/, is the director of the Indiana University Center for Bioethics, associate dean for bioethics in the Indiana University School of Medicine; and professor of Medicine, Medical and

Molecular Genetics, Public Health and Philosophy at Indiana University. He received his PhD from the bioethics program in Philosophy at Georgetown's Kennedy Institute of Ethics. Meslin came to Indiana in 2001 from Washington, D.C., where he previously directed bioethics research for the Ethical, Legal and Social Implications program of the Human Genome Project, and then served as executive director of the US National Bioethics Advisory Commission established by President Bill Clinton. He has published several books and more than 150 articles and book chapters on topics ranging from international health research to science policy.

Synthetic Life: A New Industrial Revolution (2012)

GREGORY KAEBNICK

In what follows I will be talking about synthetic biology: a technology that, I think it's safe to say, relatively few people had heard of five years ago. At that time the Hastings Center was contacted by the Alfred P. Sloan Foundation to explore the possibility of running a research project generally laying out the ethical issues associated with this emerging field in hopes that we could kick start the scholarly bioethical discussion of it.

This is still a brand new field. Many in the field and many critics of it, as well as mutual observers, think that this is going to be the enabling technology of a second industrial revolution. We've been through one. We're about to begin a second in the era of biological engineering. So if that's right, then it seems to me that the overarching question about the ethical and social implications is simply whether we can do a better job the second time around than we did the first. Could we avoid making some of the mistakes? Could we maybe even clean up some of the mistakes we made the first time around, like the environmental damage, the public health harms, and the social inequalities that were created?

Some people in synthetic biology will tell you flat out that their motivation is, at least in part, to save the world. I've heard both Drew Andy and Craig Venter say something along these lines at one time or other. On the other hand, critics of the field think that we have in synthetic biology the final ruination of the world. It is a development that merely completes the problems begun in the first industrial revolution: what Prince Charles, about twelve years ago in the *Guardian* called "the Industrialization of Life" with a capital L, although he was writing at the time about agricultural biotechnology (Charles 1999).

I don't quite want to endorse the thought that this is the beginning of a second industrial revolution. I want to be agnostic about that and, quite frankly, I regard that question as above my pay grade. I'm just an ethicist and here I only comment on the kinds of questions that synthetic biology is going to raise. So I'm going to wait and see along with you what kind of revolution, if any, it is actually going to bring. But I certainly see a lot of promise in the field and, even if it's something less than the enabling technology for a new revolution, it still leaves important moral questions to be addressed.

These are not necessarily new questions. They're really versions of very old questions and they have to do with how we're going to assess the outcomes—the risks and benefits—and weigh them against each other. They challenge us to consider how we're going to think about the challenge that this technology poses, if it poses one, to moral questions about the human relationship to nature, and how to ensure social justice. However, in "synbio," these questions do present themselves in particularly challenging forms. The plan for what follows is to walk through these questions and to lay out my views on them as they are so far developed.

Current Synthetic Biology Is Metabolic Engineering

The first thing to do is just to try to get a handle, if we can, on what synbio is. It's always a good idea in bioethics to try to lay out the facts as well as you can. There is a mantra at the Hastings Center that Tom Murray, the president of the Center, is deeply fond of: that is, "good ethics begins with good facts." In synbio this is particularly crucial because some of the questions that have been raised and some of the objections that have been lodged have to do with what I think are slight misunderstandings of the field, some even promulgated within the field itself to some degree. A more careful look at the reality of the field is warranted.

Now, it happens that the definition of synbio is itself hotly contested within biology. Rodger Brent, who calls himself a genome biologist and has been a member of both projects the Hastings Center has run on synbio, has argued during our discussions that the field does not warrant a new name. It is really just an advanced form of the kind of genetic engineering that's been around for a couple of decades now. Synbio may be quantitatively different—there are better tools, more information, more automation—but it's not qualitatively different. It's not a different kind of activity altogether. Others within the field, like Drew Endy and Craig Venter, are equally firm that it is an entirely new way of thinking about biology (Endy, Bassler, and Carlson 2010; Venter, Church, and Prather 2010). The basic contrast, of course, is with analytic biology. Analytic biology was about understanding biological systems and synthetic biology is about trying to construct them.

There are a handful of definitions that have been adopted by the many reports generated on synbio in the last decade or so. The definitions all reflect a basic understanding, a core idea, that synbio involves engineering in a way that what we now can think of as "traditional" genetic engineering never managed to do. Genetic engineering up and until now has been engineering only in a kind of metaphorical sense because it was haphazard. We really didn't know enough about what we were doing. This time around, however, instead of fighting against the complexity of biological systems, we're going to take an approach that will center on simplifying, modularizing, and standardizing them: making their behavior predictable so that we can achieve sufficient mastery of their structures and use them then to build biological systems. These systems will function like computers or factories, safely producing the products we want whenever we want in the amounts that we want.

In addition to these questions about the basic understanding of what the field is, there are also a handful of different taxonomies that have been developed to explain the lines of work that fall under the heading. In a report on synthetic biology that was released at the end of 2010, the Presidential Commission for the Study of Bioethical Issues settled on a twofold taxonomy. Others have used a fourfold taxonomy. I have sometimes used a fourfold taxonomy myself, but I have recently concluded that a three-part taxonomy captures what people in the field mean when they talk about it: bio-bricks, synthetic genomics, and protocell creation.

One line, namely the line that aims most directly at this grand vision of integrating biology and engineering, is the bio-bricks line. Bio-bricks are genetic sequences that perform very specific functions, like turning gene production on or off, or measuring the concentration of a particular gene product. Then you can assemble these genetic sequences in larger sequences or systems, sometimes called circuits (borrowing terminology from electrical engineering). Once assembled, you can install them in organisms, turning those organisms into specialized tools of sorts. You can do this without necessarily knowing very much about the underlying biology. In theory, you could some-day do it on your kitchen counter: garage-bio. So there's a shift from esoteric science to engineering not only conceptually, but also professionally. A lot of people in synthetic biology, for example Drew Endy (Endy 2005), are really keen to kick start this professional development.

For synbio to be something that everybody can play at, you really need an open source intellectual framework for these biological parts. You have to have them available to the public. It would also be helpful if they were quite well characterized and offered in public catalogs of some sort. About a year ago, in an online catalog called the "MIT Registry of Standardized Biological Parts," you could see that just such a catalog was being developed. The people who have been using that catalog and contributing to it are teams of primarily undergraduate students who have participated in the International Genetically Engineered Machines Competition. At the beginning of the summer, they're given a kit of parts from the registry and then, working over the summer at their own schools with mentors, they use those parts and new parts of their own invention to produce specially engineered bugs, or more accurately more specially engineered biological systems that can inserted into larger biological systems or microbes, to do various things. One year, for example, the winner was a system that could make *E. coli* detect arsenic in well water. The idea was to have a detector that would be cheap and could be used in developing coun-tries to gauge water contamination.

The second line in the taxonomy is associated with Craig Venter. It's what he and others at the J. Craig Venter Institute, JCVI, call synthetic genomics. One of their goals is to build a simple genome that contains only the genetic material needed to sustain minimal life, basic bacterial life - the simplest pos-sible organism. Such a minimal genome could then be kind of a standardized platform into which you could insert genetic sequences such as bio-bricks.

So one of the technological steps that made synbio possible is the develop-ment of highly automated tools for synthesizing DNA sequences, something practically like a DNA printer. The price of synthesizing has come way, way down and it will eventually be available at the level of desktop computing. This line of work puts such technology front and center. It's also this line of work that really put the whole field of synbio front and center a couple of years ago and led to the Presidential Commission's report on synbio in 2010. In May of that year, scientists from JCVI had taken a step toward creating a minimal genome by successfully synthesizing the entire genome of the bacteria *Myco-plasma mycoides* (Gibson et al. 2010). As a proof of their success, they inserted the genome into the cell of a closely related species, *Mycoplasma capricolum*, resulting in what was to all appearances a fully functioning *mycoides*. They described the cell they had created as a synthetic cell, the first ever, and they also said it represented a new species, which they dubbed *Mycoplasma mycoides JCVI-syn 1.0*. Some commentators also called this a synthetic life form, the first ever (Wade 2010).

There are two reasons for skepticism about this claim that JCVI had created a synthetic species. First, most serious commentators it as a gross exaggeration (Wade 2010). They held that JCVI had synthesized a genome, not a cell. JCVI defended its claim on the ground that the cell essentially belongs to its genome, so the cell takes on the identity of the genome. But one could also easily say that the cell had adopted or taken up the genome; or, simply that the two entities, the genome and the cell, had managed to sync up or integrate with each other. Second, commentators also pointed out that JCVI had borrowed the genome's design: the result was really more the mimicry of nature than an act of creation (Wade 2010). Nonetheless, it is still a wonderful accomplishment, proof of principle that a genome could be synthesized despite not being everything that it was cracked up to be.

The third line in the taxonomy of synthetic biology might be cobbled together out of what are really several disparate lines of research, all united in that they seek to reinvent the basic mechanism and materials found in living things. For example, in what's known as protocell creation or the creation of chemical cells, or *chells*, there's work going on to design organisms from the ground up, abstracting from all existing forms of life and identifying the basic functions that are necessary for life at its most basic (Deamer 2005). Protocell creation seeks out mechanisms for metabolism, for control, for organization, and for replication (if that's also necessary for life), and attempts to construct them from raw materials.

These protocells might use chemicals not found in naturally occurring organisms. In principle, they could have an entirely new biochemistry—a nonorganic biochemistry, in that it was not carbon based. Further, they might have tools of replication that are not based on DNA, or that employed new forms of DNA.

Those are the pretty basic lines within the taxonomy of synthetic biology. In order to clarify the field of synthetic biology, however, I want to complicate that taxonomy just a little bit by offering a few examples. These examples will also help me to articulate three sets of moral questions that I want to discuss toward the end.

First, consider the development of microorganisms that give us a new way of producing biofuels. This is one of the most commonly cited potential applications for synbio. It's probably often thought to be a somewhat nearer term application and it offers something like a base case for considering ethical questions.

There are multiple lines of research under way simultaneously using different organisms, aiming at the production of different kinds of fuel. I'm just going to give an outline of a couple of lines of this research. One line is aimed at refining the metabolism of organisms so they can more efficiently produce fuel from substrates found in crops like corn, sugar cane, or sugar beets, and turn those into biofuels like ethanol or butanol. Baker's yeast and *E. coli* are two target organisms for this line of work toward ethanol production (Jang et al. 2012). The genetic modifications consist both of the insertion of foreign genes and the elimination of native ones. These tweaks aim to allow the organisms to produce ethanol more efficiently, eliminate metabolic pathways that compete with ethanol production, make them tolerate higher levels of ethanol, since ethanol is toxic to the cells, and broaden the range of substrates they can work with. In the dream scenario for this line of research, you would have cells that could process not just glucose from sugarcane but also cellulose from switch grass or woodchips, something that would be much cheaper to produce.

For butanol production, target organisms include strains of clostridia and *E. coli*. Some clostridia are natural butanol producers, and the genetic engineering consists of increasing their efficiency. Another line of research in biofuels is on developing organisms that can produce some of these same kinds of fuels photosynthetically. Target organisms here include algae and blue-green algae—cyanobacteria, which are not actually a kind of algae. Biomass from feedstock wouldn't be necessary in this case. The inputs would just be carbon dioxide, water and sunlight. The dream scenario here is that you would not only have a method of producing fuel that avoided or at least limited the harms associated with drilling and transporting oil and growing feed stocks to produce biomass, but also—since the process is itself carbon fixing—a method that would help offset some of the environmental costs of burning the fuel in vehicles later on. Of course the process is not without costs of its own. Some methods of producing fuels would require really heavy capital outlay. For instance, for what are basically greenhouses (often very optimistically portrayed) to produce fuel in any quantity, you'd need a tremendous investment up front. More likely would be the creation of organisms that could work in open ponds. But then you would need vast plantations of open ponds.

The second example is sometimes considered the flagship example for synthetic biology. But it's also kind of, as the expression goes, the camel's nose. This is the case that's going to try to justify the field morally for the rest of us: it's the production of artemisinin. Artemisinin is a drug to treat malaria, and the most effective existing treatment as I understand it. But it also happens to be fairly expensive and scarce. It's traditionally extracted from the wormwood plant, which can be grown in plantations but is not easily grown in the quantities necessary to maximize production of this drug.

The synbio approach to artemisinin production is to construct a new metabolic pathway, bringing together genes from bacteria, yeast, and wormwood, and to insert that pathway into microbes so they can produce artemisinic acid, which is a precursor to artemisinin. The finished artemisinin can then be produced in the laboratory.

One preliminary point I want to make about these cases is that they demonstrate genuine benefit, and a benefit that would be obtained in the relatively near term. They're not just starry-eyed, futuristic possibilities, something that might happen twenty or thirty years down the road. Malaria is an enormous problem; we tend to forget about it a little bit in this country, but it affects hundreds of millions of people each year and kills up to one million per year (WHO 2012). This research was a joint venture of Amyris Biotechnologies, Institute for One World Health, and U.C. Berkeley, with funding from the Gates Foundation, and it has now been licensed to a pharmaceutical company, Sanofi-Aventis, which expects to actually start making synthetically produced artemisinin in late 2012.

The second point I want to make is that, though these applications are plausible, they are still difficult. They are not sure bets by any means. Amyris has also been one of the firms that has been at the forefront of the effort to make organisms that can make biofuels. But in February of 2012, an article in Technology Review announced that Amyris was giving up most of its biofuels production. Here's a quote from that article, "the company learned firsthand just how difficult to achieve the kinds of yields seen in lab tests in large scale production. In an update call for investors, CEO John Melo is humbled by the lessons we have learned" (Bullis 2012). Instead of biofuels, the company is going to focus on higher value, smaller quantity chemicals such as moisturizers for cosmetics, which is, if your goal at the outset was to save the world, probably a little bit deflating.

The final thing I want to note about these lines of work is that none of these cases really clearly illustrates the more sensationalistic claims that I gave you in the beginning of what synthetic biology amounts to. None of these involves the creation of life or the development of platform organisms that can be outfitted with interchangeable bio-bricks. None of these is the integration of biology with engineering in a way that will allow it to be done in our kitchens. This is all still very, very arcane and difficult work that they're doing. For this kind of work, I think that the label "synthetic biology" is probably a little bit over-blown. I think that a better label might be "metabolic engineering," which some in the field are actually beginning to adopt a little bit. Jay Keasling, who is the driving force at Amyris, has been using this term recently (Nielsen and Keasling 2011).

Metabolic engineering is the study and alteration of metabolic processes within existing organisms. So this work really does resemble traditional genetic engineering. It's essentially gene transfer, just on a bigger scale. The creation of organisms that can create artemisinic acid is bringing together not just a gene or two, but suites of genes from multiple organisms. They're doing a lot all at once. But it doesn't quite fit those descriptions of synthetic biology that I gave you at the beginning. And again, I'm not saying this to belittle the field, but only to try to get a clearer view about what synbio is actually turning out to be.

Potential Benefits, Potential Harms

Now we're in a position to think about the ethics of this developing field, and about whether and how to go forward with it. So far, most of the debate has centered on potential benefits and potential harms. The cases I've discussed have already made the potential benefits clear enough, I hope.

However, there is also great concern about potential harms. At the first meeting of a project that the Hastings Center ran on synthetic biology a couple of years ago, there was a presentation by a prominent scientist—a very enthusiastic, pro-technology guy whom I will not name, since he was speaking under conditions of confidentiality during the project—who said that the only thing that worried him about that this developing technology was that it might lead to catastrophe. The 1918 Spanish flu virus has been briefly recreated in the laboratory. The polio virus has been recreated. Eventually it's going to be possible recreate many more bacterial pathogens, and it will be possible not just to recreate existing pathogens but to redesign them, to modify them to make the more virulent, and maybe to come up with something entirely new.

In one way this threat is entirely plausible—in fact, it's already basically been demonstrated by recreating some of these viruses—but at the same time the seriousness of the threat is debated. For one thing, just creating a pathogen is only the first step in launching anything like a bioterrorist strike. You also have to weaponize the pathogen, and that process, in itself, poses a big technical challenge. Then someone would have to deliver it: a big logistical as well as technical challenge. As we learned in World War I gas attacks, such weapons can't really be aimed really well. It is of the nature of pandemics that they work their way around the globe. So such synthetic pathogens might strike the people who had launched them, and conceivably they could impact those countries harder than the intended targets.

It's also worth remembering that not every kind of application of synthetic biology raises this concern about bioterror in just the same way. Of course almost any research in the field might lead to misuse indirectly, but some applications, like the development of organisms to produce fuel, don't directly pose that concern. More serious for that case study is a second category of concern, which waggish commentators have sometimes dubbed "bio-error," to contrast with bioterror (Caruso 2008). The concern here is basically about accidents. It's that synthetic organisms will escape from confinement and their genetic sequences will wreak havoc in the environment or with public health or with agriculture. There are various ways this could happen. Modified organisms might escape and simply display unexpected properties in their new environment. Or perhaps they might mutate to acquire them. Or maybe they would trade genetic sequences with organisms in the environment, either by trading some portion of their own genome to those organisms or by acquiring something from an organism out there. That kind of lateral transfer is very common with microorganisms. So in one way or another, you might end up with an organism that has new and undesirable properties.

Here again, the seriousness of the threat is debated. For one thing, these hypothetical synthetic biologists would have likely created a simplified stripped-down organism that would be less likely to have the genetic flexibility and the system redundancy that would allow it to cope well with changing environmental challenges. Also, organisms that were committed to producing excess quantities of ethanol would very probably be at a considerable evolutionary disadvantage compared to the wild organisms that they encountered in the field. In fact, many in synbio will tell you that keeping synthetic organisms separated from the environment is more important to protect those organisms from the environment than it is to protect the environment from those organisms.

In addition to these passive reasons for reassurance, there might be some active measures that could be taken to limit bio-error and the resulting harms.

For example, there's been a lot of talk in synthetic biology about fail-safe mechanisms that could be implemented to keep an organism from surviving once it escaped into the field. There is also the possibility of physical containment. If you had organisms producing biofuel within secured greenhouses or some sort of glass trays, they are going to be physically confined to that facility, which provides an added level of safety.

Needless to say, none of these reassurances or protections has quelled concern—and none of it really should quell concern. Living organisms are complex adaptive systems, and so there is inherently a high level of unpredictability around what they're going to be capable of and how they're going to function in the environment. As it was pointed by an ecologist at the hearings of the Presidential Commission (Snow, Thomas, and King 2010), once the organism escapes into the environment it would be very difficult to eradicate—probably even impossible to eradicate. Environmental contamination by microbes would be different from contamination by chemical spills.

Compare these potential risks of synbio to the explosion of the Deep Water Horizon in 2010. The point of this thought experiment is twofold. On the one hand, it just serves to underscore the point that accidents will happen, and fail-safe mechanisms will fail. We should not be too easily reassured about the safety of the field. We should take a hard look at it. But, on the other hand, it also serves in a way to call attention to some of the benefits of the field. If we were able to contribute to our fuel supply by means of synthetic organisms, maybe, then, there would be somewhat less need for extracting oil from the Gulf of Mexico, and disasters like the 2010 spill would be less frequent.

The Way Forward

So what's the way forward concerning risks and benefits? At the outset, we need very careful research on the biology and the ecology of the "synthetic" organisms that have been proposed for commercial application. We also need information about how they reproduce, how they spread, whether and how they exchange genes with other strains, whether they might become more abundant or die out, rates of evolution, what kinds of side effects do they produce (Dana, Kuiken, Rejeski, and Snow 2012).

Everyone agrees that we also need to review our oversight and regulatory mechanisms. We need to make sure they are appropriate for the structure of the field, whatever that turns out to be. They have to be flexible mechanisms because the field is constantly evolving. So if you come up with something that is keyed specifically to the technological capabilities that are currently in the field, it is going to become almost immediately obsolete. We need a system that also functions globally in some ways since threats can emerge anywhere and are not going to observe national boundaries. And at the same time that it's very broad and reaching, you need an oversight and regulatory system that is going to be relatively fine-grained, if you will, given the possibility that the field will turn out more like engineering than like an esoteric science. If it goes that way, important even innovative work could be done in small-scale labs, which could be very difficult to monitor and regulate. We need to have an approach to handle those concerns.

In addition, I think we also need to give more thought to the underlying psychological and philosophical questions about the understanding of potential outcomes and how they're evaluated. These are questions about what counts as risk or a benefit, how heavily to weight particular risks, including catastrophic risk, and how much to discount a risk or potential benefit that is low in probability or would only occur years down the road.

We also need to think more seriously about what it means to take a so-called "precautionary stance" to a new biotechnology. The *precautionary principle* is an approach to thinking about risks and potential benefits that was developed by environmentalists as a strategy for heading off unexpected environmental damage. The basic idea in it is to shift the burden of proof from those who raise concerns about some new industrial project to those who are promoting that project. Instead of waiting for proof from opponents that the project will cause some damage, society can ask those proponents pushing the project to prove it won't cause harm.

Philosophers and proponents of emerging biotechnologies are fairly widely agreed that at least certain crude formulations of the precautionary principle simply don't work on the grounds that they lead to a kind of developmental paralysis: we can never sufficiently prove that a project won't cause harm. This is how the counter argument goes: if we're forced to always keep looking for a "we don't know what yet," then we'll end up with nothing but indeterminable testing and second-guessing, and we'll miss out on all the potential benefits.

In its 2011 report the Presidential Commission tried to strike a compromise of sorts between the precautionary principle and what it called a proactionary stance. The Commission argued for what it called "prudent vigilance" to guide the assessment of outcomes. If the precautionary principle is something like watching and waiting, prudent vigilance is something like watching but not waiting—going forward while monitoring things carefully. In the eyes of critics, however, the Presidential Commission ended up with a position that tilted very heavily toward the proactionary end of things. The report basically supports going forward on every application that it considered, although it did call for additional testing before field release. So it seems to be calling for caution and oversight and more public discussion, but not really raising any actual barriers to development.

In March 2012, a loose consortium of those critics collaborated on a response statement that tried to articulate a rather robustly precautionary stance (Friends of the Earth, ETC Group, and International Center for Technology Assessment 2012). This statement calls for a broad moratorium on applications using organisms developed through synthetic biology. By my reading this statement goes a little too far in the precautionary direction. It so happens this consortium is adopting a version of the Precautionary Principle that is more sophisticated than the really crude sort I described earlier, but nonetheless tilts somewhat too strongly toward a precautionary stance, by my lights. The statement says, for example, that before we go forward with synbio, all the alternative approaches have to be "fully considered first" (Friends of the Earth, ETC Group, and International Center for Technology Assessment 2012, 3). Of course, there is no way of specifying what all the alternative approaches are or what it means to fully consider them. This particular requirement tends to function a little more like a flat prohibition than a precaution.

The document also says that the precautionary approach requires synthetic biology-specific oversight mechanisms that account for the unique characteristics of synthetic organisms and their products. I have a hard time figuring out exactly what is going on here. It is certainly true that particular synthetic organisms would be engineered to have unique properties, but that alone wouldn't create a requirement for specific oversight mechanisms. There are going to be many drugs that have unique properties, but not every drug with a unique property has an oversight mechanism unique to that drug. The claim here seems to be more that synthetic organisms as a class are physically different from other organisms. They have unique properties as a class, simply because they've been genetically modified. But this is flatly false. You might actually create an organism using these techniques that was identical to existing organisms, and in fact that's very nearly what JCVI managed to do when it created the synthetic mycoides cell. That cell has a different history to it, but the actual physical properties of that organism as it is sitting there are not necessarily unique. What is really needed is that we pay attention to the characteristics of particular modified organisms and the plausible, if not fully established, threats and harms associated with them.

What underlies the very strong requirements in this new statement from these critics is, I think, another kind of objection altogether that is lying more or less hidden inside it. This objection is that there is something intrinsically bad about developing synthetic organisms using synthetic biology techniques. These are not merely organisms developed using new techniques; these are synthetic organisms, synthetic life forms, which to some people look like a kind of bogeyman. Now, you might find such labels attractive—and the people who came up with the phrase "synthetic biology" do find those labels attractive. To them, the term illustrates knowledge, creativity, and industry, things that we value in and of themselves. When you're at synthetic biology conferences, it's fascinating to hear synthetic biologists themselves talk about what they think is valuable in their work. They will mention the benefits, maybe something about saving the world, but normally they do it because they are really into the challenge of it. For them, it's the learning and the building; it's a kind of humanistic enterprise, really, at bottom. It's human industry, the advancement of human knowledge, and advancement of human mastery over the world. But the critics look at the same work and they see something that is intrinsically morally troubling. I say "morally troubling" in order to capture a broader set of possible positions that would be capture if we were talking only about morally right and wrong, if we were only talking about permissibility.

One potential source of intrinsic concern about synthetic biology is that it might prove that some cherished views about life are just flatly false. When JCVI announced its synthetic cell, some claimed that science had finally shown once and for all that life is just a well-organized puddle of chemicals (Caplan 2010). There is no greater being nor any spiritual core inside living things in virtue of which life has any sacred properties. But I think this conclusion is mistaken—and this is actually the other reason for remaining skeptical about "synthetic life." What JCVI created was a synthetic genome, not a cell. But let's suppose it was a full synthetic cell. If there is a God who gives microbes in the wild some special vital essence, then it would also be within that God's powers to endow synthetic microbes, microbes created within the lab, with those same properties as well.

A similar question arose at an earlier presidential bioethics commission, the National Bioethics Advisory Commission, when it considered cloning. The question arose during NBAC's meetings whether human clones would have souls. I think it is extremely difficult to think they would not, however, if other people also have souls. Similarly, it seems to me that microbes created in the labs could have whatever special soul-like properties that characterize microbes generally.

So even if we manage to create synthetic organisms, we needn't say that we've shown life is something other than what we thought it was.

Maybe the classic way of understanding intrinsic concerns about synbio is kind of allied to this question that synbio shows life to be merely a puddle of chemicals. The classic formulation is that synbio and genetic engineering involves a kind of religious or metaphysical mistake. Rather than proving that there is nothing of God in living things, as the earlier objection I just walked through charges, the idea here is that we're invading God's turf. This would be maybe the most natural way, on the face of it, of understanding the phrase "playing God."

But one problem in understanding this moral argument is that if the religious or metaphysical claims are really quite robust, then they're not likely to be very widely shared. They're not likely to have very much traction in public debate. As such, it will be, I think, a little problematic to invoke them in public debate as ground for public policy. As it happens, however, most religious organizations so far have been fine with what's going on in synthetic biology. After JCVI announced its "synthetic cell," the Vatican issued a statement praising the work (BBC Monitoring Europe 2010).

I actually think that this claim is not what most people have in mind even when they use the phrase "playing God." Mostly they're just trying to find a way of articulating a rather slippery moral point.

Drawing on literature in environmental ethics, you might say there is a general moral choice in our relationship to nature between two kinds of discourses or competing moral ideals governing the relationship between humans and nature: one of alternating nature to meet human demands, and the other of adjusting human demands to accommodate nature. The first holds that nature is just stuff to be used, pumped out of the ground, cut down, burned, turned to waste, disposed of. The second calls on us to limit the harm that we are doing to nature.

I like this distinction. But at the same time, we can think of various cases in which the distinction, and the application of the distinction, is not going to be very clear-cut at all. My favorite example is gardening. Science writer Michael Pollan has written a lot about the philosophy of gardening, if you will. In his book *Second Nature: A Gardener's Education*, he writes that "nature abhors a garden" (Pollan 1991, 37). But at the same time, as Pollan also stresses, a successful garden is distinctly allied with nature. A garden is simultaneously a case of altering nature to meet human demands and of altering human demands to accommodate nature—particularly if one of the reasons that we garden is that we think that by growing some of our own food we're eating really locally and reducing our environmental impact.

I think you could say similar things of the greener ways of producing energy—developing solar power or wind power, for example. So I find the distinction illuminating. But my point is that it isn't always a sharp distinction. Synbio is somewhat like solar power, somewhere in the middle between these two discourses. If we focus on the grand definitions of synbio, then the field looks like the example of adjusting nature to meet demands. But if we focus on some of the applications of it, the concrete forms the field is taking, then I think it looks less threatening and more accommodating. If algae were modified to make gasoline, for example, that would really be a more dramatic way of adapting nature to meet humans demands than drilling for oil, shipping it around the globe, processing it in huge refineries, and so on. Arguably, I think, synbio might likewise be seen as adapting human ends to nature—as a "green" technology. Now there is a question as to whether synbio is really greener. This turns our attention from the intrinsic issue of genetic engineering to the effect on the environment.

One real life example to add to the hypothetical example of biofuel. Hard cheeses are produced using an enzyme called chymosin, which is found in rennet, which is in turn produced from the stomachs of un-weaned calves. It turns out that it is possible to skip the calf-stomach step: you can take cow genes and insert those into microbes so that they will produce the chymosin directly. This has been around for quite a while now. It was introduced in the early 1990s and approved for commercial use. By 2008, 80% to 90% of commercially made cheese in the United States and the UK were made using chymosin from genetically modified organisms. The question for us is: would using chymosin from modified microbes be a clearer case of adapting nature to human ends than using rennet from slaughtered calves' stomachs? It seems to me that it's not.

The third concern I'm going to touch on has to do with the social distribution of benefits and harms. In a nutshell, the benefits of synthetic biology will likely accrue to wealthy nations at the expense of developing nations. They will be excluded or even actively exploited. The case of bioproduction illustrates these concerns quite sharply. Producing ethanol from sugar in sugarcane, for example, would require growing incredible amounts of sugarcane. Producing biofuel from algae would require vast, vast arrays of ponds. Either of these might turn out to be harmful to the environment, and if the land would otherwise be used for growing food for the people in those countries, then it would be directly bad for those people too. On the other hand,

as the Presidential Commission for the Study of Bioethical Issues emphasized, part of the appeal of synbio in the first place is this "saving the world"
feature: it could help address the long standing significant problem of social
and economic injustice. The case of artemisinin illustrates that. So how can
we achieve the vision?

Final Questions

In general, with respect to all three of the questions I've identified, how do we
make the technology turn out for the best? I'll leave you with a few outstanding
questions that I think need more study.

First, how do we go about deliberating publically about these questions
and about the outcomes of this field? There is a lot of fine-sounding talk from
almost everyone who offers ethical commentary on synthetic biology about the
importance of public deliberation. But how do you actually do that? How do
you make sure that the public is represented accurately in that process? And
how do you make sure that the field is represented accurately in that process,
that the facts are correct?

Second, how do you go about steering the development of a technology
toward one set of outcomes rather than another? Can we do that? Or does
technological development fundamentally depend on a kind of undirected
creativity?

Third, who bears responsibility for ensuring good outcomes? Do governments representing society as a whole or do private citizens and corporations
that are involved in synthetic biology? Do all of us also individually shoulder
some responsibility? The Presidential Commission suggested that responsibility should be shared by both private citizens and also corporations active in
the field. I worry that by distributing responsibility this way we are making it
a little too diffuse.

I have a compost bucket on my kitchen counter at home. That is what
passes for the use of microorganisms in my house and is probably as close as
I am likely to get to kitchen-counter manipulation of microbes. I tell you this
sort of to lay my cards on the table and to indicate to you what my general
biases are. I am drawn toward the discourse of adjusting human demands to
accommodate nature. I'm not innately a proponent of synthetic biology. I don't
find it as inherently attractive in the same way that those at the professional
conferences do. But, my bottom line is that I think the critics are mistaken to
find synbio objectionable across the board. The question to me is if there are
better and worse ways of developing and using this technology. How can we get

ahead of potential problems and not just solve them as they occur? An answer to this question calls for a new and focused way of thinking about technological development.

References and Recommended Reading

BBC Monitoring Europe. 2010. "Vatican Dismisses Synthetic Cell's Life-Giving Dimensions, Lauds Science Research." *BBC Monitoring International Reports*, May 25.

Bullis, Kevin. 2012. "Amyris Gives Up Making Biofuels: Update." *Technology Review*. http://www.technologyreview.com/view/426866/amyris-gives-up-making-biofuels-update/.

Caplan, Arthur. 2010. "Now Ain't That Special? The Implications of Creating the First Synthetic Bacteria." *Scientific American*, May 20. http://blogs.scientificamerican.com/guest-blog/2010/05/20/now-aint-that-special-the-implications-of-creating-the-first-synthetic-bacteria/.

Caruso, Denise. 2008. *Synthetic Biology: An Overview and Recommendations for Anticipating and Addressing Emerging Risks*. Washington, D.C.: Center for American Progress.

Charles, Prince of Wales. 1999. "Questions about genetically modified organisms." *Daily Mail*, June 1.

Dana, Genya V., Todd Kuiken, David Rejeski, and Allison A. Snow. 2012. "Synthetic Biology: Four Steps to Avoid a Synthetic-Biology Disaster." *Nature* 483 (7387): 29. doi:10.1038/483029a.

Deamer, David. 2005. "A Giant Step Towards Artificial Life?" *Trends in Biotechnology* 23 (7): 336–38. doi:10.1016/j.tibtech.2005.05.008.

Endy, Drew. 2005. "Foundations for Engineering Biology." *Nature* 438 (7067): 449–53. doi:10.1038/nature04342.

Endy, Drew, Bonnie L. Bassler, and Robert Carlson. 2010. "Overview and Context of the Science and Technology of Synthetic Biology." Presentation to the First Meeting of the Presidential Commission for the Study of Bioethical Issues, July 8–9, 2011. Washington, D.C. http://bioethics.gov/cms/node/164.

Friends of the Earth, ETC Group, and International Center for Technology Assessment. 2012. *The Principles for the Oversight of Synthetic Biology*. Washington, D.C.

Gibson, Daniel G., John I. Glass, Carole Lartigue, Vladimir N. Noskov, Ray-Yuan Chuang, Mikkel A. Algire, Gwynedd A. Benders, et al. 2010. "Creation of a Bacterial Cell Controlled by a Chemically Synthesized Genome." *Science* 329 (5987): 52–56. doi:10.1126/science.1190719.

Jang, Yu-Sin, Jong Myoung Park, Sol Choi, Yong Jun Choi, Do Young Seung, Jung Hee Cho, and Sang Yup Lee. 2012. "Engineering of Microorganisms for the Production of Biofuels and Perspectives Based on Systems Metabolic Engineering Approaches." *Biotechnology Advances* 30 (5): 989–1000. doi:10.1016/j.biotechadv.2011.08.015.

Kaebnick, Gregory E. 2001. "On Genetic Engineering and the Idea of the Sacred: A Secular Argument." *St. Thomas Law Review* 13 (4): 863–76.

Kaebnick, Gregory E. 2005. "Behavioral Genetics and Moral Responsibility." In *Wrestling with Behavioral Genetics: Science, Ethics, and Public Conversation*, edited by Erik Parens, Audrey Chapman, and Nancy Press. Baltimore: Johns Hopkins University Press.

Kaebnick, Gregory E. 2007a. "Putting Concerns About Nature in Context: The Case of Agricultural Biotechnology." *Perspectives in Biology and Medicine* 50 (4): 572–84. doi:10.1353/pbm.2007.0049.

Kaebnick, Gregory E. 2007b. "Shaping Our Future: Law, Policy, and Ethics." In *An Era of Reproductive Genetics*, edited by Lori Knowles and Gregory E. Kaebnick. Baltimore: Johns Hopkins Press.

Kaebnick, Gregory E. 2008. "Reasons of the Heart: Emotion, Rationality, and the 'Wisdom of Repugnance.'" *Hastings Center Report* 38 (4): 36–45. doi:10.1353/hcr.0.0037.

Kaebnick, Gregory E. 2009. "Should Moral Objections to Synthetic Biology Affect Public Policy?" *Nature Biotechnology* 27 (12): 1106–8. doi:10.1038/nbt1209-1106.

Nielsen, Jens, and Jay D. Keasling. 2011. "Synergies Between Synthetic Biology and Metabolic Engineering." *Nature Biotechnology* 29 (8): 693–95. doi:10.1038/nbt.1937.

Pollan, Michael. *Second Nature: A Gardener's Education*. New York: Grove Press, 1991.

Snow, Allison, Jim Thomas, and Nancy M. P. King. 2010. "Benefits and Risks." Presentation to the Presidential Commission for the Study of Bioethical Issues, July 8–9, 2011. Washington, D.C. http://bioethics.gov/cms/node/166.

Venter, Craig, George Church, and Kristala L. J. Prather. 2010. "Applications." Presentation to the first meeting of the Presidential Commission for the Study of Bioethical Issues, July 8–9, 2011. Washington, D.C. http://bioethics.gov/cms/node/165.

Wade, Nicholas. 2010. "Researchers Say They Created a 'Synthetic Cell.'" *The New York Times*, May 20.

World Health Organization (WHO). 2012. *Malaria*. http://www.who.int/topics/malaria/en/.

Biographical Sketch

Gregory E. Kaebnick, http://www.thehastingscenter.org/About/Staff/Detail.aspx?id=1286, is a research scholar, the director of the editorial department, and editor of the Hastings Center Report and Bioethics Forum. He received his PhD in philosophy from the University of Minnesota in 1998. Kaebnick's research focuses on the intersections between value and emerging biotechnologies. Most recently in this context, he has worked on appeals to nature and on issues in synthetic biology.

INDEX